The Spinal Cord during the First and Early Second Trimesters 4- to 108-mm Crown-Rump Lengths

This fourteenth of 15 short atlases reimagines the classic 5 volume *Atlas of Human Central Nervous System Development*. This volume presents sections of the spinal cord from specimens between 4 mm and 108 mm with detailed annotations. Extensive 3-D reconstructions show the early development of the germinal zones, the earliest neurons and early white matter accumulations at cervical levels only. Other 3-D reconstructions in older specimens show the progressive segregation of motor neuron columns at all levels in the ventral gray.

The Glossary (available separately) gives definitions for all the terms used in this volume and all the others in the *Atlas*.

Key Features

- Classic anatomical atlases
- Detailed labeling of structures in the developing spinal cord offers updated terminology and the identification of unique developmental features, such as germinal matrices of specific neuronal populations and migratory streams of young neurons
- Appeals to neuroanatomists, developmental biologists, and clinical practitioners
- A valuable reference work on brain development that will be relevant for decades

ATLAS OF
HUMAN CENTRAL NERVOUS SYSTEM DEVELOPMENT
Series

Volume 1: The Human Brain during the First Trimester 3.5- to 4.5-mm Crown-Rump Lengths

Volume 2: The Human Brain during the First Trimester 6.3- to 10.5-mm Crown-Rump Lengths

Volume 3: The Human Brain during the First Trimester 15- to 18-mm Crown-Rump Lengths

Volume 4: The Human Brain during the First Trimester 21- to 23-mm Crown-Rump Lengths

Volume 5: The Human Brain during the First Trimester 31- to 33-mm Crown-Rump Lengths

Volume 6: The Human Brain during the First Trimester 40- to 42-mm Crown-Rump Lengths

Volume 7: The Human Brain during the First Trimester 57- to 60-mm Crown-Rump Lengths

Volume 8: The Human Brain during the Second Trimester 96- to 150-mm Crown-Rump Lengths

Volume 9: The Human Brain during the Second Trimester 160- to 170-mm Crown-Rump Lengths

Volume 10: The Human Brain during the Second Trimester 190- to 210-mm Crown-Rump Lengths

Volume 11: The Human Brain during the Third Trimester 225- to 235-mm Crown-Rump Lengths

Volume 12: The Human Brain during the Third Trimester 260- to 270-mm Crown-Rump Lengths

Volume 13: The Human Brain during the Third Trimester 310- to 350-mm Crown-Rump Lengths

Volume 14: The Spinal Cord during the First and Early Second Trimesters 4- to 108-mm Crown-Rump Lengths

Volume 15: The Spinal Cord during the Middle Second Trimester through the 4th Postnatal Month 130- to 440-mm Crown-Rump Lengths

The Spinal Cord during the First and Early Second Trimesters 4- to 108-mm Crown-Rump Lengths

Atlas of Human Central Nervous System Development, Volume 14

Shirley A. Bayer

Joseph Altman

CRC Press

Taylor & Francis Group

Boca Raton London New York

CRC Press is an imprint of the
Taylor & Francis Group, an **informa** business

Designed cover: Shirley A. Bayer and Joseph Altman

First edition published 2024
by CRC Press
2385 NW Executive Center Drive, Suite 320, Boca Raton FL 33431

and by CRC Press
4 Park Square, Milton Park, Abingdon, Oxon, OX14 4RN

CRC Press is an imprint of Taylor & Francis Group, LLC

LCCN no. 2022008216

ISBN: 978-1-032-22905-8 (hbk)
ISBN: 978-1-032-22904-1 (pbk)
ISBN: 978-1-003-27470-4 (ebk)

DOI: 10.1201/9781003274704

Typeset in Times Roman
by KnowledgeWorks Global Ltd.

Publisher's note: This book has been prepared from camera-ready copy provided by the authors.
Access the Support Material: www.routledge.com/9781032229058

CONTENTS

ACKNOWLEDGEMENTS

We thank the late Dr. William DeMyer, pediatric neurologist at Indiana University Medical Center, for access to his personal library on human CNS development. We also thank the staff of the National Museum of Health and Medicine that was at the Armed Forces Institute of Pathology, Walter Reed Hospital, Washington, D.C. when we collected data in 1995 and 1996: Dr. Adrianne Noe, Director; Archibald J. Fobbs, Curator of the Yakovlev Collection; Elizabeth C. Lockett; and William Discher. We are most grateful to the late Dr. James M. Petras at the Walter Reed Institute of Research who made his darkroom facilities available so that we could develop all the photomicrographs on location rather than in our laboratory in Indiana. Finally, we thank Chuck Crumly, Neha Bhatt, Kara Roberts, Michele Dimont, and Rebecca Condit for expert help during production of the manuscript.

AUTHORS

Shirley A. Bayer received her PhD from Purdue University in 1974 and spent most of her scientific career working with Joseph Altman. She was a professor of biology at Indiana-Purdue University in Indianapolis for several years, where she taught courses in human anatomy and developmental neurobiology while continuing to do research in brain development. Her lengthy publication record of dozens of peer-reviewed, scientific journal articles extends back to the mid 1970s. She has co-authored several books and many articles with her late spouse, Joseph Altman. It was her research (published in *Science* in 1982) that proved that new neurons are added to granule cells in the dentate gyrus during adult life, a unique neuronal population that grows. That paper stimulated interest in the dormant field of adult neurogenesis.

Joseph Altman, now deceased, was born in Hungary and migrated with his family via Germany and Australia to the US. In New York, he became a graduate student in psychology in the laboratory of Hans-Lukas Teuber, earning a PhD in 1959 from New York University. He was a postdoctoral fellow at Columbia University, and later joined the faculty at the Massachusetts Institute of Technology. In 1968, he accepted a position as a professor of biology at Purdue University. During his career, he collaborated closely with Shirley A. Bayer. From the early 1960s-2016, he published many articles in peer-reviewed journals, books, monographs, and free online books that emphasized developmental processes in brain anatomy and function. His most important discovery was adult neurogenesis, the creation of new neurons in the adult brain. This discovery was made in the early 1960s while he was based at MIT, but was largely ignored in favor of the prevailing dogma that neurogenesis is limited to prenatal development. After Dr. Bayer's paper proved new neurons are added to granule cells in the hippocampus, Dr. Altman's monumental discovery became more accepted. During the 1990s, new researchers "rediscovered" and confirmed his original finding. Adult neurogenesis has recently been proven to occur in the dentate gyrus, olfactory bulb, and striatum through the measurement of Carbon-14—the levels of which changed during nuclear bomb testing throughout the 20th century—in postmortem human brains. Today, many laboratories around the world are continuing to study the importance of adult neurogenesis in brain function. In 2011, Dr. Altman was awarded the Prince of Asturias Award, an annual prize given in Spain by the Prince of Asturias Foundation to individuals, entities, or organizations globally who make notable achievements in the sciences, humanities, and public affairs. In 2012, he received the International Prize for Biology - an annual award from the Japan Society for the Promotion of Science (JSPS) for "outstanding contribution to the advancement of research in fundamental biology." This Prize is one of the most prestigious honors a scientist can receive. When Dr. Altman died in 2016, Dr. Bayer continued the work they started over 50 years ago. In her late husband's honor, she created the Altman Prize, awarded each year by JSPS to an outstanding young researcher in developmental neuroscience.

INTRODUCTION

ORGANIZATION OF THE ATLAS

This is the 14th volume in a series of Atlases on the development of the human central nervous system. This volume deals with the development of the spinal cord in the first and early second trimesters. These specimens were presented in Volume 1 of the original Atlas Series (Bayer and Altman, 2001). Parts I to IV feature photographs of transversely cut spinal cords from normal specimens ranging in age from gestational week (GW) 3.2 through GW 14 with crown-rump lengths (CR) from 4 to 108 mm. The specimens are drawn from three collections housed in the National Museum of Health and Medicine in Silver Springs, MD.[1, 2, 3]

Each specimen or set of specimens is introduced by an *overview plate* that shows thumbnail photographs of all sections in that part of the Atlas at the same scale. The overview plate is followed by *companion plates* designated as **A** and **B** on facing pages. The **A** part of each plate on the left page shows the full contrast high-magnification photograph[4] of the specimen without any labels; the **B** part of each plate on the right page shows a low-contrast copy of the same photographs with superimposed outlines and unabbreviated labels. **Part II** features spinal cords at the cervical level only during the early-to-middle first trimester (CR 4.0–32 mm, GW 3.2–9.6). **Parts III** to **V** feature all levels of the spinal cord in three specimens: M2050 (CR 36 mm, GW 10), Y380-62 (CR 56 mm, GW 11.9), and Y68-65 (CR 108 mm, GW 14).

Part VI features 3-D reconstructions of the cervical area in eight specimens from CR 4 to CR 11.9, following a five-step procedure. *First*, photographs of sections at regularly spaced intervals through the cervical spinal cord were scanned and converted to computer files). *Second*, the files were aligned to each other using the middle dorsoventral part of the spinal canal as the reference point. *Third*, Adobe Illustrator was used to outline 14 structures in each section.[5] *Fourth*, the files were imported into 3-D space (x, y, and z coordinates) using Cinema4DXL (C4D, Maxon Computer, Inc.), a modeling and animation software package. For each section, points on the outlines have unique x-y coordinates and share the same z coordinate. By calculating the distance between sections, the entire array of outlines is stretched out in the z axis. The outlines are segregated into 14 different groups, one for each structure.[5] The C4D loft tool builds a "skin" for each structure by creating a spline mesh of polygons. The polygons start from the x-y points on the first outline with the most anterior z coordinate, to the x-y points on the next outline behind it, and finish with the x-y points on the last outline at the most posterior z coordinate. The polygons are rendered either as completely opaque or as partly transparent surfaces using the C4D ray-tracing engine. Selected parts of the model can be made either invisible or visible during rendering using the various options in C4D. *Fifth*, the rendered images are converted to Photoshop files, and Illustrator is used to draw thin lines on some of the surfaces to make the images easier to understand.

---

1. The *Carnegie Collection* (designated by a **C** prefix in the specimen number) started in the Department of Embryology of the Carnegie Institution of Washington. It was led by Franklin P. Mall (1862–1917), George L. Streeter (1873–1948), and George W. Corner (1889–1981). These specimens were collected during a span of 40 to 50 years and were histologically prepared with a variety of fixatives, embedding media, cutting planes, and histological stains. Early analyses of specimens were published in the early 1900s in *Contributions to Embryology, The Carnegie Institute of Washington* (now archived in the Smithsonian Libraries). O'Rahilly and Müller (1987, 1994) have given overviews of some first trimester specimens in this collection.

2. The *Minot Collection* (designated by an **M** prefix in the specimen number) is the work of Dr. Charles Sedgwick Minot (1852–1914), an embryologist at Harvard University. Throughout his career, Minot collected about 1900 embryos from a variety of species. The 100 human embryos in the group were probably acquired between 1900 and 1910. From our examination of these specimens and their similar appearance, we assume that they are preserved in the same way, although we could not find any records describing fixation procedures. The slides contain information on section numbers, section thickness (6 µm or 10 µm), and stain (aluminum cochineal).

3. The *Yakovlev Collection* (designated by a **Y** prefix in the specimen number) is the work of Dr. Paul Ivan Yakovlev (1894–1983), a neurologist affiliated with Harvard University and the AFIP. Throughout his career, Yakovlev collected many diseased and normal human brains. He invented a giant microtome that was capable of sectioning entire human brains. Later, he became interested in the developing brain and col-

lected many human brains during the second and third trimesters. The normal brains in the developmental group were cataloged by Haleem (1990) and were examined by us during 1996 and 1997 when we spent time at the AFIP.

4. All sections were photographed using either an Olympus photomicroscope (the Carnegie and Minot specimens) or a Wild photomakroskop (the Yakovlev specimens) using Kodak technical pan black-and-white negative film #TP442. The film was developed for 6-7 minutes in dilution F of Kodak HC–110 developer, followed by stop bath for 30 seconds, Kodak fixer for 5 minutes, Kodak hypo clearing agent for 1 minute, running water rinse for 10 minutes, and a brief rinse in Kodak photo–flo before drying. Each specimen was photographed at the magnification that filled the microscopic field with the largest cross section of the spinal cord. Negatives were scanned at 2700 dots per inch (dpi) using a Nikon coolscan–1000 35-mm film scanner interfaced to a Macintosh PowerMac G3 computer. Normal-contrast Adobe Photoshop files are in **Part A** of each plate; low-contrast copies are in part **B** with labels inserted using Adobe Illustrator.

5. The 14 structures are: (1) the outside edge of the entire section, (2) the entire neuroepithelium on the left side, (3-5) the dorsal, intermediate, and ventral parts of the neuroepithelium on the right side, (6-7) the roof and floor plates, (8) the undivided gray matter on the left side, (9-11) the dorsal horn, intermediate gray, and ventral horn on the right side, (12) the dorsal funiculus on both sides, (13) the lateral funiculus on both sides, (14) the ventral funiculus on both sides and the ventral commissure. Since structural outlines are usually symmetrical in most specimens, the outlines from one side are copied to the other side to simplify the images. The outlines from each section are saved in separate Adobe Illustrator eps (encapsulated postscript) files.

Part VII contains 3-D reconstructions of motor columns in the entire spinal cord, using the same procedures as Part VI for three specimens in the late first trimester and early second trimester (GW 10, GW 11.9, and GW 14). Five structural sets[6] were reconstructed. The images were rendered so that the white and gray matter appear as transparent, glass-like surfaces surrounding the opaque surfaces of the central canal and the motoneuron columns. Because the spinal cord is many times longer than it is wide, the thin motoneuron columns are difficult to see when all axes are rendered at the same scale, especially in the GW 14 model. Consequently, **the x-y axis is three times larger than the z axis in all models**.

SPINAL CORD ANATOMY

Figure 1 shows the gross structure of the spinal cord in a neonate. The spinal cord is a continuation of the brain medulla that extends through the spinal canal created by the bony structures of the individual vertebrae. Its posterior (dorsal) surface is contacted by the dorsal roots of the spinal nerves. Each spinal nerve has a swelling in the dorsal root (dorsal root ganglion) that contains the cell bodies of primary sensory neurons whose endings are embedded in the skin, viscera, and muscles. Axons from dorsal root ganglion cells enter the dorsal funiculus of the spinal cord and travel to various sensory areas of the brain as well as send collaterals into the gray matter of the spinal cord. Each spinal nerve splits and contacts the anterior (ventral) part of the spinal cord. The ventral roots contain axons of ventral horn motor neurons that control skeletal muscles. Ventral root branches in thoracic and upper lumbar areas have ganglia in the sympathetic chain; here are the cell bodies of the postganglionic autonomic motor neurons; their axons control smooth muscles and glands throughout the viscera.

The labels in this Atlas are based on a review of the literature (Altman and Bayer, 2001) on experimental studies of anatomical projections in the spinal cord of various mammals. In addition, several atlases of the adult human spinal cord are used (Fix, 1987; Roberts *et al.*, 1987; DeArmond *et al.*, 1989; Haines, 2000). For the most part, the labels follow the terminology in these atlases, especially in the gray matter. However, some parts of the white matter are labeled differently from the standard adult atlases in two ways. First, additional subdivisions are made in the dorsal funiculus, and second, the propriospinal tract is limited to a circumferential tract around the ventral horn and is renamed the intraspinal tract. Finally, we briefly explain where our developmental terminology departs from some of terms used in descriptive embryology (see also Altman and Bayer, 1981, 1984, 1995).

New Subdivisions in the Dorsal Funiculus. The largest components of the dorsal funiculus contain the principal ascending axons of dorsal root ganglion cells in the fasciculus gracilis and the fasciculus cuneatus. However, experimental evidence indicates that the entry zone of the dorsal root fibers and the fibers bordering the medial side of the dorsal horn have a more complex organization. As the dorsal root axons enter, they first bifurcate into ascending and descending branches (see Chapter 4, Section 4.1.2 in Altman and Bayer, 2001). That part of the dorsal funiculus is labeled the *dorsal root bifurcation zone*. Indeed, that part of the dorsal funiculus is the first to develop, appearing as an indistinct structure around GW 5.2 (**Plate 4**) and more definite swelling in the white matter by GW 5.25 (**Plate 5**). In the developmental literature, that structure is called the oval bundle of His, and our labels in the youngest specimens give both names to the structure.

After the dorsal root fibers bifurcate, there is developmental evidence in rats (Kudo and Yamada, 1987) that both the ascending and descending branches give off short, local collaterals (see Chapter 4, Section 4.2.2 in Altman and Bayer, 2001). These collaterals hug the medial edge of the dorsal horn before they enter the gray matter. For that reason, that part of the dorsal funiculus is called the *dorsal root collateralization zone*. Some of the collaterals invade the top of the dorsal horn in a dorsal bundle that first grows down, then curves upwards to form elaborate terminal arbors in various laminae of the dorsal horn. Other collaterals enter the dorsal horn in a dorsomedial bundle that grows ventrally and forms arbors within the motoneuron columns in the ventral horn. Just like the dorsal root bifurcation zone, the dorsal root collateralization zone is an early developing part of the dorsal funiculus. It first appears around GW 7.75 (**Plate 9**) and GW 7.8 (**Plate 10**).

The latest developing and eventually the largest component of the dorsal funiculus contains the principal ascending axons of the dorsal root ganglia in the traditionally named fasciculus gracilis and fasciculus cuneatus. Most probably, the ascending branch that initially bifurcates and gives off local collaterals in the lateral part of the dorsal funiculus near its point of entry shifts medially as it grows toward the brain to enter these fasciculi. We call this a shift from "local" to "express" lanes (see Chapter 4, Section 4.2.3 in Altman and Bayer, 2001). These fasciculi are set off from the rest of the dorsal funiculus by shallow depressions on either side of the dorsal midline. The first evidence that these fasciculi appear is on GW 9.6 (**Plates 12, 14–20**).

The Intraspinal (Propriospinal) Tract. This tract, which carries local connections between neurons in the spinal cord that do not ascend to the brain, is traditionally called the *proprio*spinal tract. The prefix *proprio-* means "within itself." Unfortunately, that prefix can be confused with a set of *proprio*ceptive fibers that transmit sensory information from muscles and tendons. By that criterion, the spinocerebellar tracts along the margin of the lateral

---

6. The five structures are: (1) the outside edge of the entire section, (2) the gray matter on both sides, (3) the central canal, (4) motoneuron columns in the right ventral horn, (5) motoneuron columns in the left ventral horn.

Posterior/Dorsal View of the Brain and Spinal Cord in a Newborn Infant

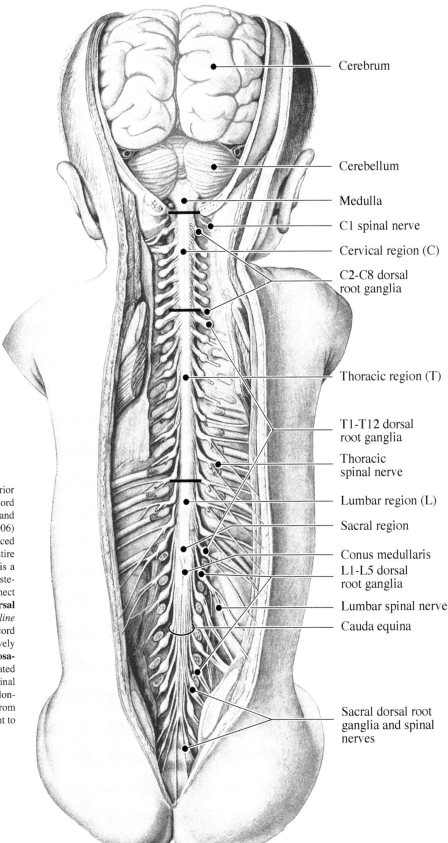

Cerebrum

Cerebellum

Medulla

C1 spinal nerve

Cervical region (C)

C2-C8 dorsal root ganglia

Thoracic region (T)

T1-T12 dorsal root ganglia

Thoracic spinal nerve

Lumbar region (L)

Sacral region

Conus medullaris

L1-L5 dorsal root ganglia

Lumbar spinal nerve

Cauda equina

Sacral dorsal root ganglia and spinal nerves

Figure 1. Dissection exposing the posterior aspect of the brain (*yellow*) and spinal cord in a neonate, drawn by Johannes Sobotta and published in the first German edition (1906) of his *Atlas of Human Anatomy* (reproduced from the 8th English edition, 1963). The entire dorsal/posterior aspect of the **spinal cord** is a continuation of the brain medulla. The posterior/dorsal roots of the **spinal nerves** connect to lateral parts. All spinal nerves have a **dorsal root ganglion** except C1. The *top thick line* indicates the border between the spinal cord and the medulla; *lower thick lines* successively segregate the **cervical, thoracic,** and **lumbosacral** regions. The **conus medullaris** is located just below the coccygeal terminus of the spinal cord, and the **cauda equina** contains the elongated roots of the spinal nerves that exit from the spine far below their point of attachment to the spinal cord.

funiculus are also propriospinal tracts because they send proprioceptive information to the cerebellum. To avoid that confusion, we use the prefix *intra* to label this tract. There is additional confusion about the extent of this tract. All of the adult atlases that we consulted label that part of the dorsal funiculus that we call the dorsal root collateralization zone as part of the propriospinal (or spinospinal) tract (Fix,1987; Roberts *et al.*, 1987; DeArmond *et al.*, 1989; Haines, 2000). However, Brodal (1981) limits the propriospinal tract to a zone immediately surrounding the ventral horn, lateral intermediate gray, and lateral dorsal horn (*see* Figure 2-3B, p. 63 in Brodal, 1981) and does not include any part of the dorsal funiculus. The experimental anatomical evidence that Brodal cites to delimit the propriospinal tract is a study by Anderson (1963) in cats. In order to identify intrinsic axons in the white matter, Anderson isolated the spinal cord at first and second thoracic levels. The results indicated surviving axons were very dense in an area that surrounds the ventral horn, but were in the areas surrounding the rest of the gray matter, including the dorsal funiculus. Anderson (1963) specifically mentions that the axons remaining in the dorsal funiculus are collaterals of the dorsal roots that are still intact within the isolated segment of the spinal cord (see comment on p. 302 in Anderson, 1963). The very dense accumulation of surviving axons around the ventral horn (the yellow outlined area in Figure 1-89 in Altman and Bayer, 2001) coincides with one of the earliest myelinating regions in the ventral and lateral funiculi. It is for that reason that we label the area around the ventral horn as the principal component of the intraspinal tract. The sequence of myelination in this tract can be followed in the myelin-stained sections in Volume 15.

DEVELOPMENTAL HIGHLIGHTS

The important developmental events that occur during the early to middle first trimester include areal changes in different parts of the spinal cord (**Figure 2**) and the presumed times of stem cell generation of neurons, migration, and the onset of migration (**Figure 3**). The simplicity of spinal cord anatomy readily allows quantification of these phenomena. Unfortunately, the complexity of brain development makes quantification difficult to impossible. That is why quantification is emphasized in this volume and the final Volume 15 as opposed to the other volumes in this Atlas Series.

AREAL CHANGES IN THE HUMAN SPINAL CORD

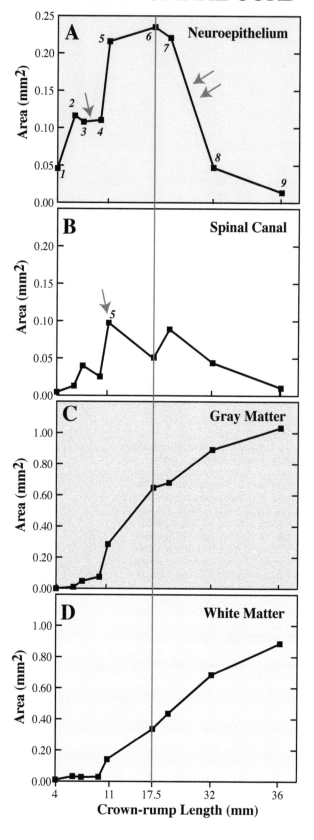

Developmental Events in Neuronal Populations during the Early to Middle First Trimester

NAME	CR 4–6.3	CR 6.7–8	CR 11–14.5	CR 17.1–32	CR 36 on
Alpha Motoneurons	Stem cells proliferate First neurons generated	Most neurons generated	Migration and settling	Segregation into columns and maturation	
Ventral Horn Interneurons	Stem cells proliferate		Neurons generated	Migration and settling	Maturation
Intermediate Gray	Stem cells proliferate		Neurons generated	Migration and settling	Maturation
Dorsal Horn Layers I,IV,V	Stem cells proliferate		Neurons generated	Migration and settling	Maturation
Dorsal Horn Layers II,III	Stem cells dormant?	Stem cells proliferate	Neurons generated	Migration and settling	Maturation
Central Autonomic Area	Stem cells dormant?	Stem cells proliferate	Neurons generated	Migration and settling	Maturation

Figure 3. The major events that take place in neuronal populations in the cervical spinal cord during the first trimester. *Dark green* indicates the first step in development when the stem cells rapidly proliferate just prior to neurogenesis. *Light green* indicates the approximate period of neurogenesis. *Pink* indicates the time during which neurons migrate away from the neuroepithelium and settle in the gray matter. *Orange* indicates the period of maturation, which continues throughout the second and third trimesters and well after birth. *Light brown* indicates the postulated time when the stem cells of neuronal populations that are generated late are either dormant or have not yet appeared. The time periods for the various events in each population are based on the data in **Figure 2** and qualitative changes visible in **Plates 2-12**.

Figure 2. The total area of the neuroepithelium (**A**), spinal canal (**B**), gray matter (**C**), and white matter (**D**) at cervical levels during the early-to-middle first trimester. The *italicized numbers* in some graphs indicate data points. The first six data points in each graph are from Minot specimens only because the same fixative was used and there is a more constant cutting plane between specimens.

The *red line* marks the peak area of the neuroepithelium (**A**) and divides developmental events. **Before the peak**, the ventral neuroepithelium expands (*data points 1 and 2*) and reaches a plateau (*single red arrow*, **A**) between *data points 2–4*. That plateau occurs at the time when the area of the ventral neuroepithelium remains stable during the production of the large motoneurons in the ventral horn (*see* **Plates 4-5**). The most rapid growth spurt in the neuroepithelium occurs between *data points 4 and 5* when the spinal canal reaches its peak area (*single red arrow*, **B**) as the dorsal neuroepithelium rapidly expands about the dorsal part of the spinal canal. There is a simultaneous growth spurt in the area of the gray matter (**C**) as neurons accumulate in the ventral horn and in the area of the white matter (**D**) as axons accumulate in the ventral funiculus, ventral commissure, and dorsal root bifurcation zone of the dorsal funiculus. These growth spurts are qualitatively visible by looking at the changing appearance of the spinal cord in **Plates 5, 6, 7**. After CR 11, the area of the neuroepithelium grows more slowly to reach a peak at CR 17.5. That slow growth is due to concurrent shrinking of the ventral neuroepithelium and expansion of the dorsal neuroepithelium (*see* **Plates 8-10**).

After the peak, neuroepithelial area gradually declines (*data point 7*) then plummets by *data point 8*. During the rapid decline (*double red arrows*, **A**; *see* **Plates 11-12**) the dorsal neuroepithelium shrinks dramatically as late-generated neurons migrate into the dorsal horn and the dorsal and ventral parts of the spinal canal disappear to leave only a central canal (**B**). During the decline in the neuroepithelium, the gray matter (**C**) and white matter (**D**) continue to increase. However, throughout the entire first trimester, gray matter area exceeds that of the white matter.

REFERENCES

Altman, J., and S. A. Bayer (1981) *Development of the Cranial Nerve Ganglia and Related Nuclei in the Rat.* (Advances in Anatomy Embryology and Cell Biology, Vol. 74.) Berlin: Springer-Verlag.

Altman, J., and S. A. Bayer (1984) *Development of the Rat Spinal Cord.* (Advances in Anatomy Embryology and Cell Biology, Vol. 85.) Berlin: Springer-Verlag.

Altman, J., and S. A. Bayer (1995) *Atlas of Prenatal Rat Brain Development.* Boca Raton, FL: CRC Press.

Altman, J., and S. A. Bayer (2001) *Development of the Human Spinal Cord. An Interpretation Based on Experimental Studies in Animals.* New York: Oxford University Press.

Anderson, F. D. (1963) The structure of a chronically isolated segment of the cat spinal cord. *Journal of Comparative Neurology*, 120:297-315.

Bayer, S. A. and J. Altman (1995) Neurogenesis and neuronal migration. In: G. Paxinos (ed.), *The Rat Nervous System*, (2nd edition) pp. 1041-1078. San Diego, CA: Academic Press.

Bayer SA, Altman J (2002) *Atlas of Human Central Nervous System Development*, Volume 1: *The Spinal Cord from Gestational Week 4 to the 4th Postnatal Month.* Boca Raton, FL, CRC Press.

Brodal, A. (1981) *Neurological Anatomy in Relation to Clinical Medicine*, (3rd edition). New York: Oxford University Press.

Corner, G. W. (1929) A well-preserved human embryo of 10 somites. *Carnegie Institution of Washington, Contributions to Embryology*, 20:81-102.

DeArmond, S. J., M. M. Fusco, and M. M. Dewey (1989) *Structure of the Human Brain A Photographic Atlas*, (3rd edition). New York: Oxford University Press.

Fix, J. D. (1987) *Atlas of the Human Brain and Spinal Cord.* Gaithersburg, MD: Aspen Publishers.

Haines, D. (2000) *Neuroanatomy: An Atlas of Structures, Sections, and Systems*, (5th edition). Philadelphia, PA: Lippincott Williams & Wilkins.

Haleem, M. (1990) Diagnostic Categories of the Yakovlev Collection of Normal and Pathological Anatomy and Development of the Brain. Washington, DC: Armed Forces Institute of Pathology.

Kudo, N., and T. Yamada (1987) Morphological and physiological studies of the development of the monosynaptic reflex pathway in the rat lumbar spinal cord. *Journal of Physiology*, 389:441-459.

Minot, C. S. (1903) *A Laboratory Text-Book of Embryology.* Philadelphia, PA: Blakiston.

Roberts, M., J. Hanaway, and D. K. Morest (1987) *Atlas of the Human Brain in Section*, (2nd edition). Philadelphia, PA: Lea & Febiger.

Sobotta, J. (1963) *Atlas of Human Anatomy*, (8th English Edition). (Volume III, Part II: *Atlas of Neuroanatomy. Central Nervous System, Autonomic Nervous System Eye, Ear, and Skin*, F. H. J. Figge, ed.) New York: Hafner.

Streeter, G. L., C. H. Heuser, and G. W. Corner (1951) Developmental horizons in human embryos: Description of age groups XIX, XX, XXI, XXII, and XXIII, being the fifth issue of a survey of the Carnegie Collection. *Carnegie Institution of Washington, Contributions to Embryology*, 34:165-196.

PART II:
Cervical Levels of the Early to Middle First Trimester
CR 4–32-mm, GW 3.2–9.6

Plate 1 is a survey of sections from the cervical levels of the spinal cord in the first 11 specimens. All sections are shown at the same scale. The boxes enclosing each section list the gestational week (GW), the crown-rump length (CR) in millimeters (mm), the specimen number preceded by M (Minot Collection) or C (Carnegie Collection), the slide number and section number from the set of slides containing all the sections of that specimen, and the total area of the section in square millimeters (mm^2). Since all Minot specimens have consecutive section numbers, no slide number is given. Full-page normal contrast photographs of each specimen are in **Plates 2A–12A.** Low-contrast photographs with superimposed labels and outlines of section details are in **Plates 2B–12B.** All A-B plates are on facing pages to allow the user to glance back and forth at the unlabeled structure on the left and the labeled one on the right. Plates **2–12** also feature high-magnification views of the spinal cord alone paired with low-magnification views of the spinal cord surrounded by other parts of the embryo.

During early stages of spinal cord development, a detailed study of the cervical level alone provides an overview of the most dramatic changes that take place during spinal cord development. It is during this period that the spinal cord changes its composition from a prominent germinal matrix (the neuroepithelium) with no continuous gray matter and a thin surrounding layer of primordial white matter (GW 3.2, specimen M714) to a small core of an ependymal/glial matrix surrounded by a large gray matter, which is in turn surrounded by a large white matter (GW 9.6, specimen C609). The gray matter at GW 9.6 contains the same populations of neurons in the ventral horn, intermediate gray, and dorsal horn as the adult spinal cord. **The period of neurogenesis is completed during the first trimester.**

The first neurons outside the ventral neuroepithelium at GW 3.2 are the earliest generated motoneurons in the ventral horn. The earliest generated neurons in the intermediate gray and the dorsal horn are outside the neuroepithelium at GW 5.2, while the population of ventral horn motoneurons continues to increase. Between GW 6 and GW 6.75, the ventral horn grows rapidly as the ventral neuroepithelium declines. Already by GW 6.6, there is a prominent depression in the center of the ventral neuroepithelium (**Plate 6B**) that may mark the location of the neuronal stem cells that gave rise to motoneurons. By GW 7.8, the ventral part of the neuroepithelium is transforming into an ependymal/glial proliferative matrix and is no longer a source of neurons. The intermediate part of the neuroepithelium expands in preparation for the growth of the intermediate gray between GW 6.6 (especially specimen M2161) and GW 8.3. By GW 9.6, the intermediate neuroepithelium has transformed into an ependymal/glial matrix. Up to GW 6.6 (especially specimen C8998), the dorsal part of the neuroepithelium is only a small component of the total neuroepithelium. There is a dramatic increase in its length between GW 6.6 and GW 7.3. That lengthening is characterized by basal "undulations" that may represent accumulations of premigratory postmitotic neurons, the postulated "sojourn zones" that have been well documented with [^{3}H]thymidine autoradiography in the spinal cord of rats on embryonic day 15 (*see* Chapter 3, Figure 3-10 in Altman and Bayer, 2001). These undulations are still present on GW 9.6 indicating that a few neurons are still being generated by the dorsal neuroepithelium. The most dramatic growth of the dorsal horn occurs between GW 8.6 and GW 9.6, accompanied by a sharp decrease in the size of the dorsal neuroepithelium during that same time. Many of these specimens are shown in the companion reference book, *Development of the Human Spinal Cord* (Altman and Bayer, 2001). Specifically, changes in the neuroepithelium and the growth of the gray matter is discussed in Chapter 5, Section 5.3.

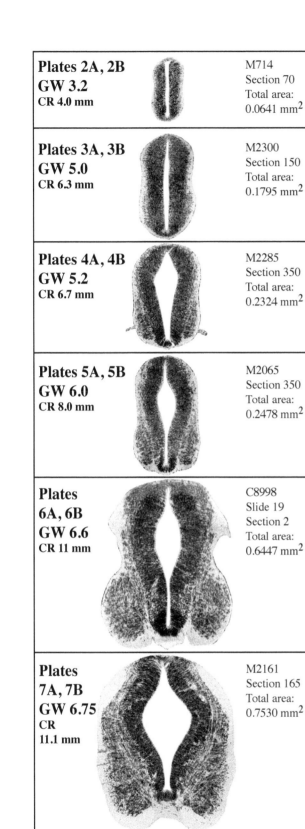

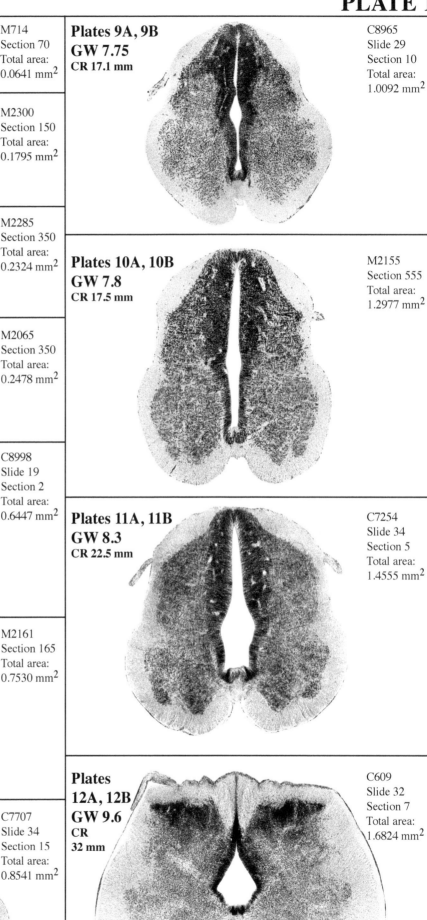

Plates 2A, 2B
GW 3.2
CR 4.0 mm

M714
Section 70
Total area:
0.0641 mm^2

Plates 3A, 3B
GW 5.0
CR 6.3 mm

M2300
Section 150
Total area:
0.1795 mm^2

Plates 4A, 4B
GW 5.2
CR 6.7 mm

M2285
Section 350
Total area:
0.2324 mm^2

Plates 5A, 5B
GW 6.0
CR 8.0 mm

M2065
Section 350
Total area:
0.2478 mm^2

Plates 6A, 6B
GW 6.6
CR 11 mm

C8998
Slide 19
Section 2
Total area:
0.6447 mm^2

Plates 7A, 7B
GW 6.75
CR 11.1 mm

M2161
Section 165
Total area:
0.7530 mm^2

Plates 8A, 8B
GW 7.3
CR 14.5 mm

C7707
Slide 34
Section 15
Total area:
0.8541 mm^2

0.25 mm

Plates 9A, 9B
GW 7.75
CR 17.1 mm

C8965
Slide 29
Section 10
Total area:
1.0092 mm^2

Plates 10A, 10B
GW 7.8
CR 17.5 mm

M2155
Section 555
Total area:
1.2977 mm^2

Plates 11A, 11B
GW 8.3
CR 22.5 mm

C7254
Slide 34
Section 5
Total area:
1.4555 mm^2

Plates 12A, 12B
GW 9.6
CR 32 mm

C609
Slide 32
Section 7
Total area:
1.6824 mm^2

0.25 mm

PLATE 2A

CR 4.0 mm
GW 3.2
M714
Cervical
Cell body stain

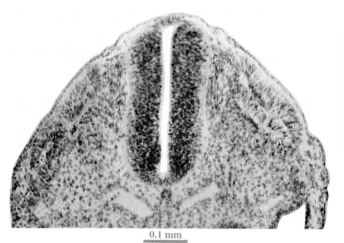

0.1 mm

Areas (mm²)	
Central canal	.0046
Neuroepithelium	.0459
Roof plate	.0021
Floor plate	.0023
Primordial white matter	.0010

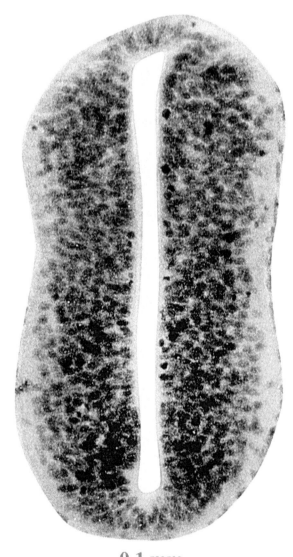

0.1 mm

PLATE 2B

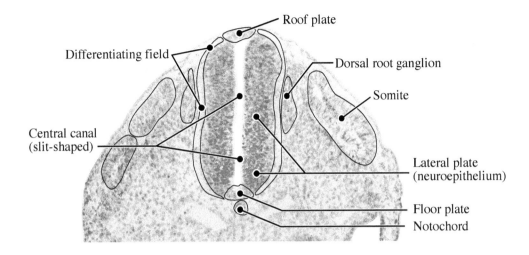

Roof plate

Differentiating field

Dorsal root ganglion

Somite

Central canal
(slit-shaped)

Lateral plate
(neuroepithelium)

Floor plate

Notochord

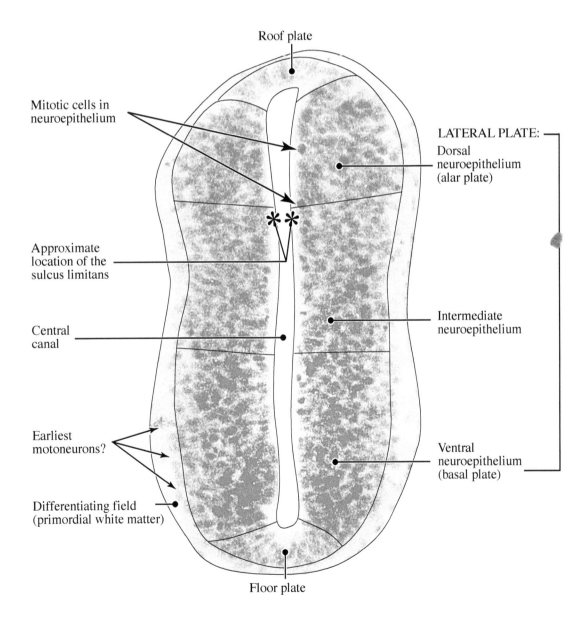

Roof plate

Mitotic cells in
neuroepithelium

LATERAL PLATE:

Dorsal
neuroepithelium
(alar plate)

Approximate
location of the
sulcus limitans

Central
canal

Intermediate
neuroepithelium

Earliest
motoneurons?

Ventral
neuroepithelium
(basal plate)

Differentiating field
(primordial white matter)

Floor plate

PLATE 3A

**CR 6.3 mm
GW 5.0
M2300
Cervical
Cell body stain**

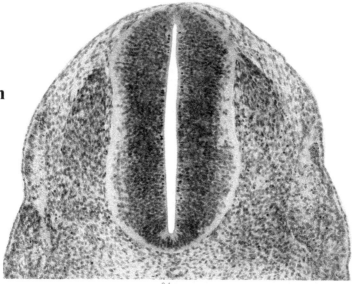

Areas (mm²)	
Central canal	.0130
Neuroepithelium	.1163
Roof plate	.0036
Floor plate	.0043
Gray matter	.0098
White matter	.0325

0.1 mm

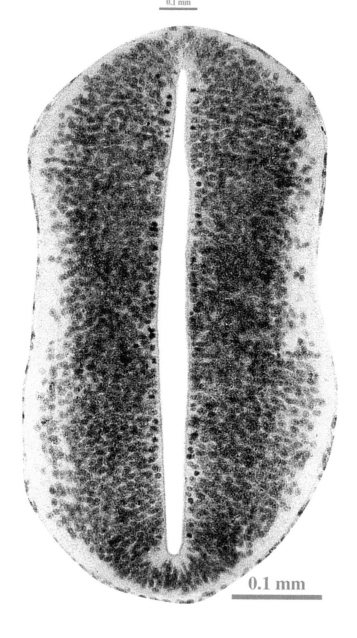

0.1 mm

PLATE 3B

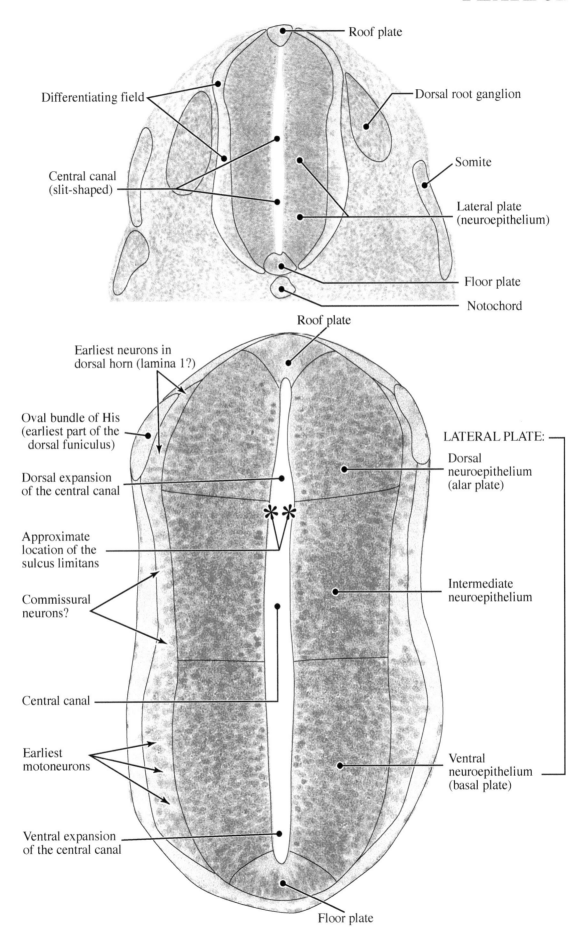

Roof plate

Differentiating field

Dorsal root ganglion

Central canal
(slit-shaped)

Somite

Lateral plate
(neuroepithelium)

Floor plate

Notochord

Roof plate

Earliest neurons in
dorsal horn (lamina 1?)

Oval bundle of His
(earliest part of the
dorsal funiculus)

Dorsal expansion
of the central canal

LATERAL PLATE:

Dorsal
neuroepithelium
(alar plate)

Approximate
location of the
sulcus limitans

Commissural
neurons?

Intermediate
neuroepithelium

Central canal

Earliest
motoneurons

Ventral
neuroepithelium
(basal plate)

Ventral expansion
of the central canal

Floor plate

14

PLATE 4A

CR 6.7 mm
GW 5.2
M2285
Cervical
Cell body stain

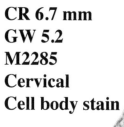

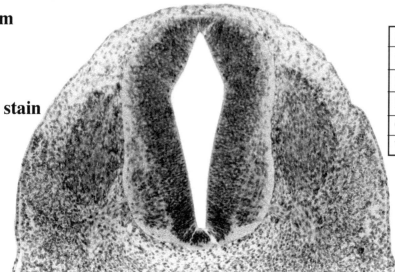

Areas (mm²)	
Central canal	.0400
Neuroepithelium	.1080
Roof plate	.0037
Floor plate	.0048
Gray matter	.0476
White matter	.2324

0.1 mm

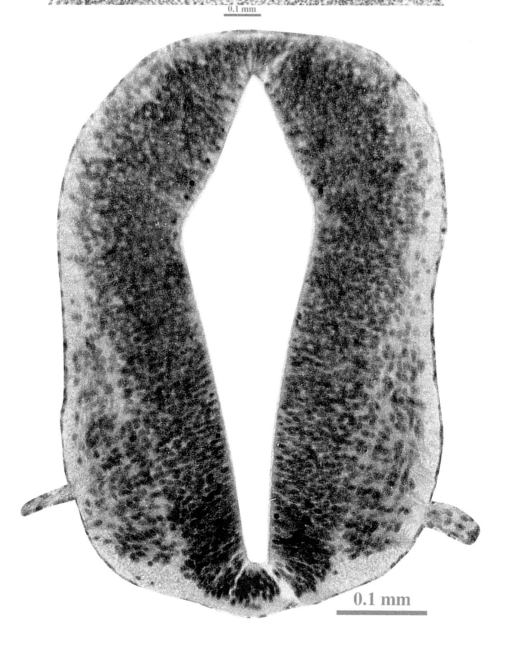

0.1 mm

PLATE 4B

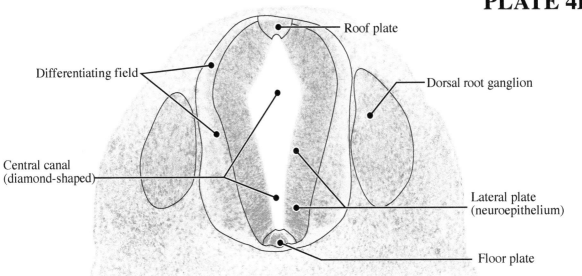

Roof plate

Differentiating field

Dorsal root ganglion

Central canal
(diamond-shaped)

Lateral plate
(neuroepithelium)

Floor plate

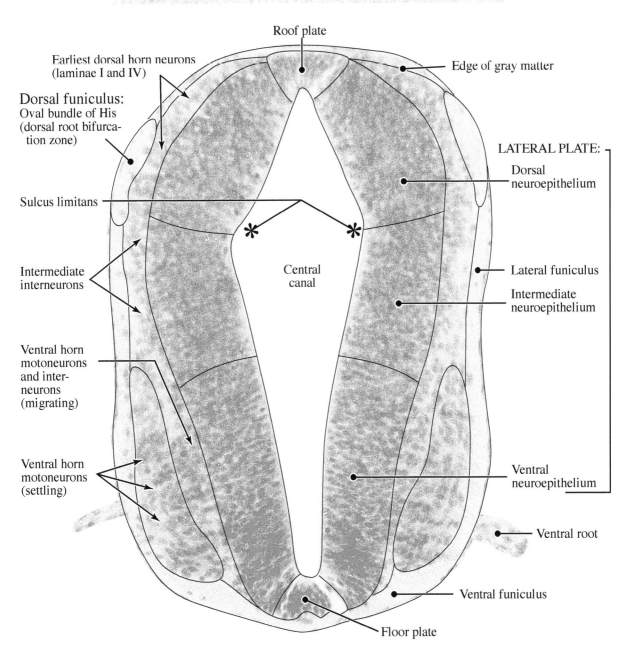

Roof plate

Earliest dorsal horn neurons
(laminae I and IV)

Edge of gray matter

Dorsal funiculus:
Oval bundle of His
(dorsal root bifurca-
tion zone)

LATERAL PLATE:

Dorsal
neuroepithelium

Sulcus limitans

✱   ✱

Intermediate
interneurons

Central
canal

Lateral funiculus

Intermediate
neuroepithelium

Ventral horn
motoneurons
and inter-
neurons
(migrating)

Ventral horn
motoneurons
(settling)

Ventral
neuroepithelium

Ventral root

Floor plate

Ventral funiculus

16

PLATE 5A

CR 8.0 mm
GW 6.0
M2065
Cervical
Cell body stain

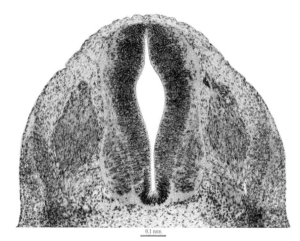

Areas (mm²)

Central canal	.0255
Neuroepithelium	.1103
Roof plate	.0012
Floor plate	.0067
Gray matter	.0752
White matter	.0289

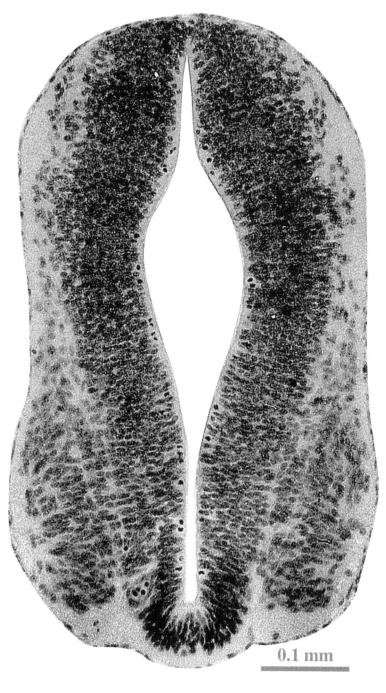

0.1 mm

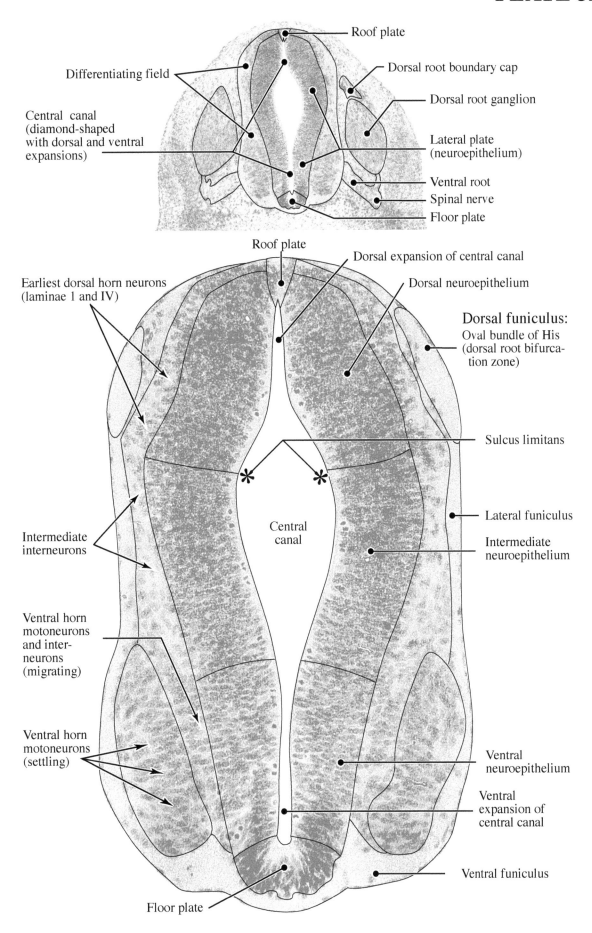

Roof plate

Differentiating field

Central canal
(diamond-shaped
with dorsal and ventral
expansions)

Dorsal root boundary cap

Dorsal root ganglion

Lateral plate
(neuroepithelium)

Ventral root

Spinal nerve

Floor plate

Roof plate

Dorsal expansion of central canal

Dorsal neuroepithelium

Earliest dorsal horn neurons
(laminae 1 and IV)

Dorsal funiculus:
Oval bundle of His
(dorsal root bifurca-
tion zone)

Sulcus limitans

Intermediate
interneurons

Central
canal

Lateral funiculus

Intermediate
neuroepithelium

Ventral horn
motoneurons
and inter-
neurons
(migrating)

Ventral horn
motoneurons
(settling)

Ventral
neuroepithelium

Ventral
expansion of
central canal

Ventral funiculus

Floor plate

18

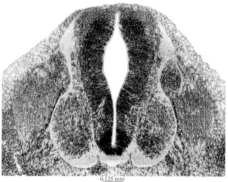

PLATE 6A

CR 11 mm
GW 6.6
C8998
Cervical
Cell body stain

Areas (mm²)	
Central canal	.0523
Neuroepithelium	.2376
Roof plate	.0029
Floor plate	.0079
Gray matter	.2522
White matter	.0919

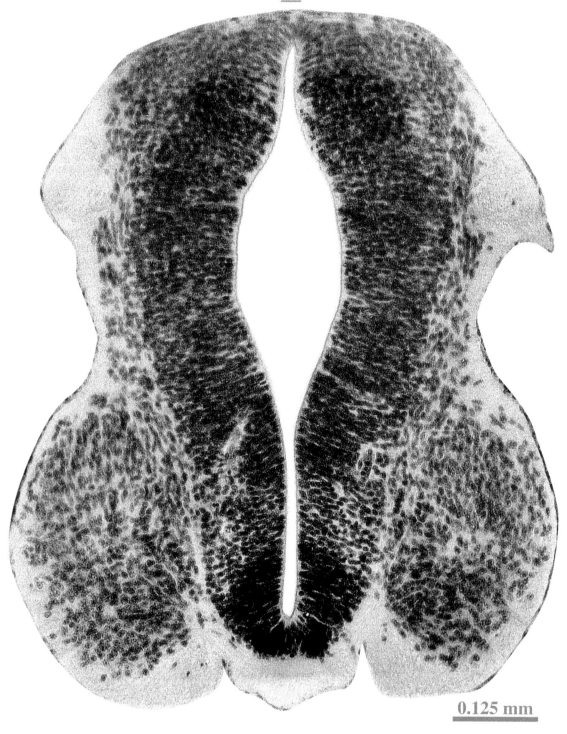

0.125 mm

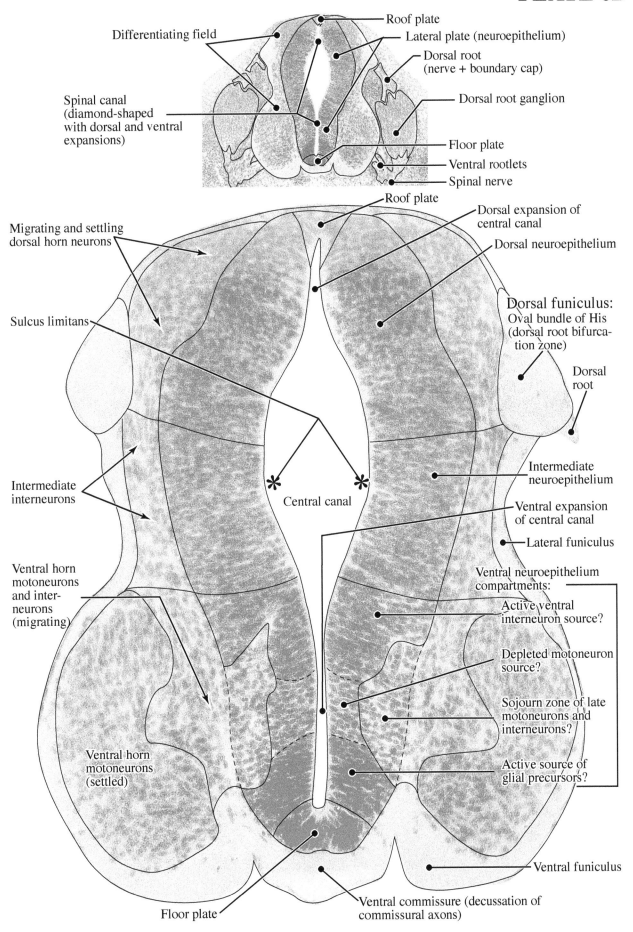

Roof plate

Differentiating field

Lateral plate (neuroepithelium)

Dorsal root
(nerve + boundary cap)

Dorsal root ganglion

Spinal canal
(diamond-shaped
with dorsal and ventral
expansions)

Floor plate

Ventral rootlets

Spinal nerve

Roof plate

Dorsal expansion of
central canal

Migrating and settling
dorsal horn neurons

Dorsal neuroepithelium

Dorsal funiculus:
Oval bundle of His
(dorsal root bifurca-
tion zone)

Sulcus limitans

Dorsal
root

Intermediate
interneurons

Intermediate
neuroepithelium

Central canal

Ventral expansion
of central canal

Lateral funiculus

Ventral neuroepithelium
compartments:

Ventral horn
motoneurons
and inter-
neurons
(migrating)

Active ventral
interneuron source?

Depleted motoneuron
source?

Sojourn zone of late
motoneurons and
interneurons?

Ventral horn
motoneurons
(settled)

Active source of
glial precursors?

Ventral funiculus

Floor plate

Ventral commissure (decussation of
commissural axons)

20

PLATE 7A

CR 11.1 mm
GW 6.75
M2161
Cervical
Cell body stain

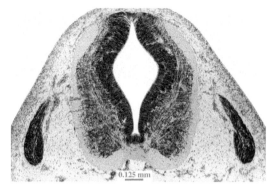

Areas (mm²)	
Central canal	.0974
Neuroepithelium	.2160
Roof plate	.0066
Floor plate	.0070
Gray matter	.2842
White matter	.1418

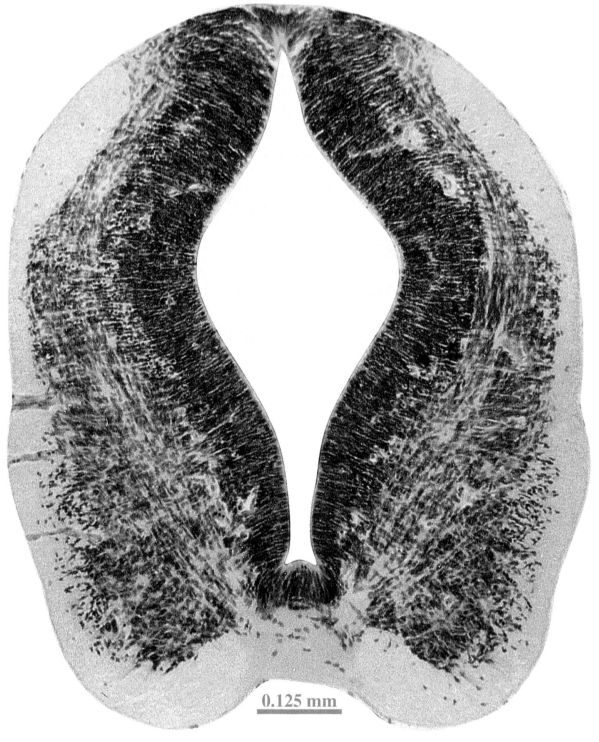

0.125 mm

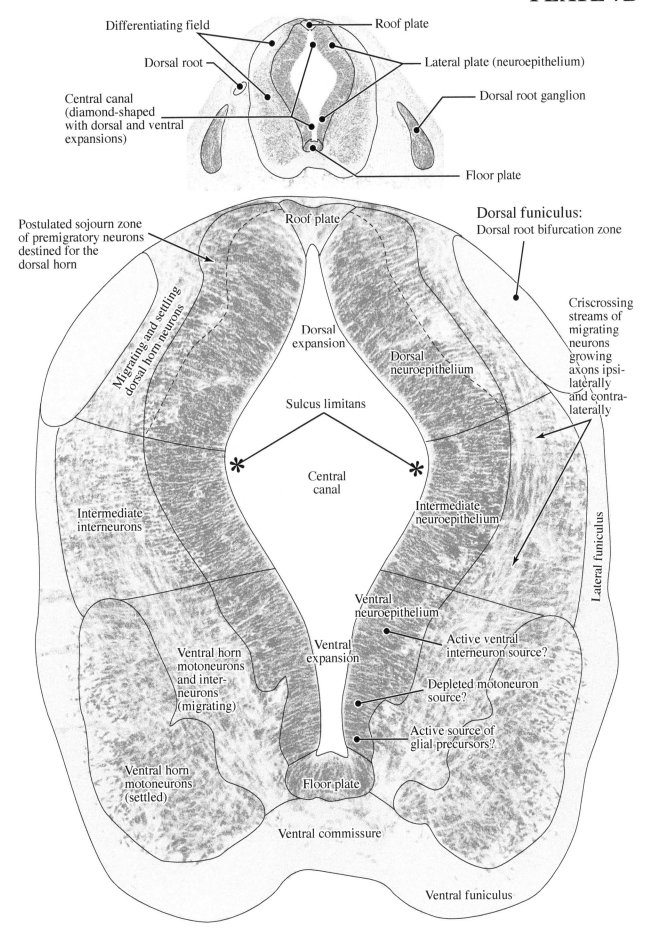

Differentiating field

Roof plate

Dorsal root

Lateral plate (neuroepithelium)

Central canal
(diamond-shaped
with dorsal and ventral
expansions)

Dorsal root ganglion

Floor plate

Postulated sojourn zone
of premigratory neurons
destined for the
dorsal horn

Roof plate

Dorsal funiculus:
Dorsal root bifurcation zone

Migrating and settling
dorsal horn neurons

Dorsal
expansion

Dorsal
neuroepithelium

Criscrossing
streams of
migrating
neurons
growing
axons ipsi-
laterally
and contra-
laterally

Sulcus limitans

Central
canal

*

*

Intermediate
interneurons

Intermediate
neuroepithelium

Lateral funiculus

Ventral
neuroepithelium

Ventral horn
motoneurons
and inter-
neurons
(migrating)

Ventral
expansion

Active ventral
interneuron source?

Depleted motoneuron
source?

Active source of
glial precursors?

Ventral horn
motoneurons
(settled)

Floor plate

Ventral commissure

Ventral funiculus

PLATE 8A

CR 14.5 mm
GW 7.3
C7707
Cervical
Cell body stain

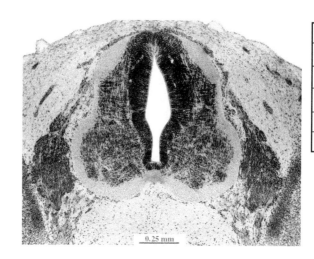

Areas (mm²)	
Central canal	.0559
Neuroepithelium	.1810
Roof plate	.0136
Floor plate	.0057
Gray matter	.3932
White matter	.2047

0.25 mm

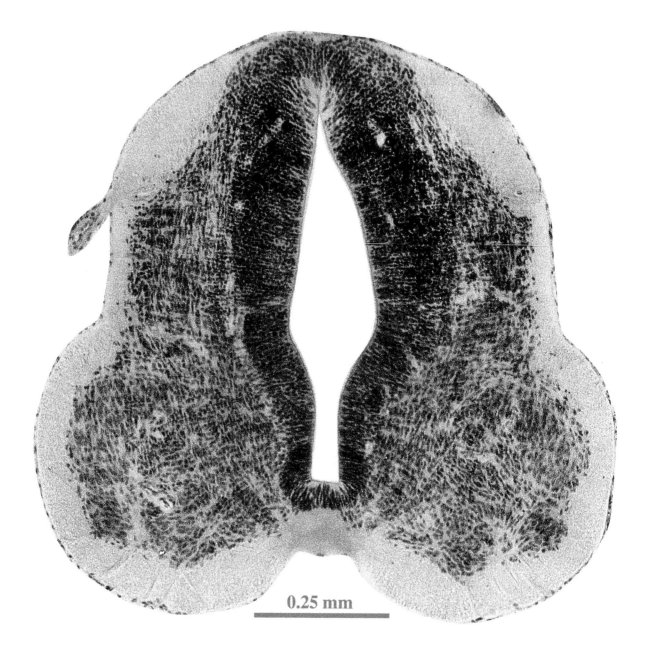

0.25 mm

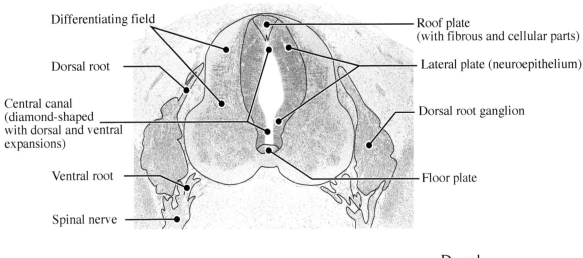

Differentiating field

Dorsal root

Central canal
(diamond-shaped
with dorsal and ventral
expansions)

Ventral root

Spinal nerve

Roof plate
(with fibrous and cellular parts)

Lateral plate (neuroepithelium)

Dorsal root ganglion

Floor plate

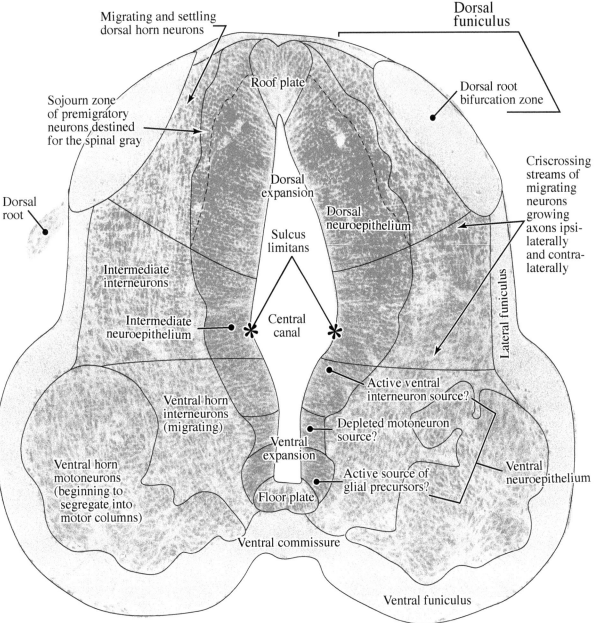

Migrating and settling
dorsal horn neurons

Dorsal
funiculus

Roof plate

Dorsal root
bifurcation zone

Sojourn zone
of premigratory
neurons destined
for the spinal gray

Dorsal
root

Intermediate
interneurons

Intermediate
neuroepithelium

Ventral horn
interneurons
(migrating)

Ventral horn
motoneurons
(beginning to
segregate into
motor columns)

Dorsal
expansion

Dorsal
neuroepithelium

Sulcus
limitans

Central
canal

∗   ∗

Ventral
expansion

Floor plate

Ventral commissure

Criscrossing
streams of
migrating
neurons
growing
axons ipsi-
laterally
and contra-
laterally

Lateral funiculus

Active ventral
interneuron source?

Depleted motoneuron
source?

Active source of
glial precursors?

Ventral
neuroepithelium

Ventral funiculus

PLATE 9A

CR 17.1 mm
GW 7.75
C8965
Cervical
Cell body stain

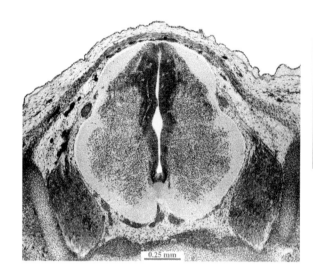

Areas (mm²)	
Central canal	.0166
Neuroepithelium	.1549
Roof plate	.0069
Floor plate	.0051
Gray matter	.5557
White matter	.2701

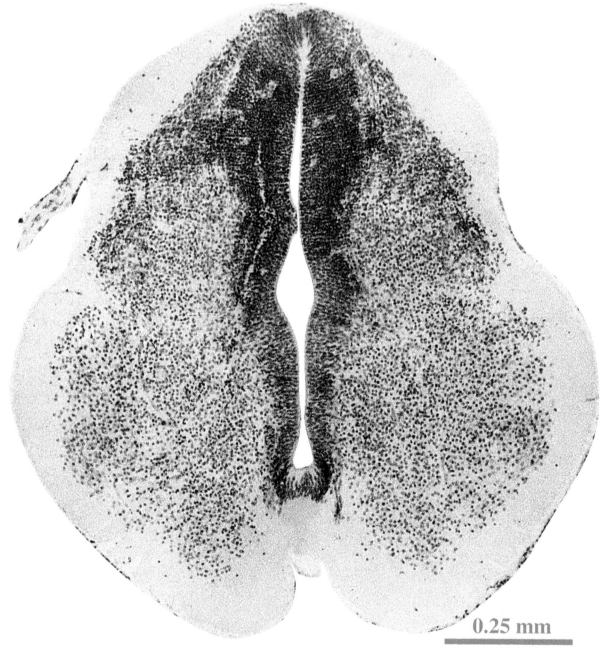

0.25 mm

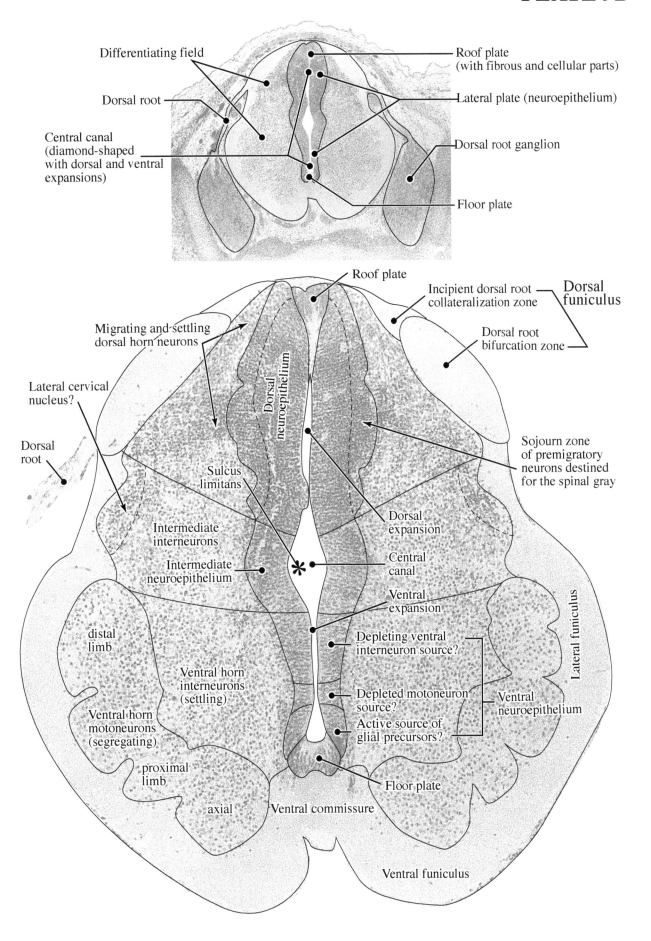

Differentiating field

Dorsal root

Central canal
(diamond-shaped
with dorsal and ventral
expansions)

Roof plate
(with fibrous and cellular parts)

Lateral plate (neuroepithelium)

Dorsal root ganglion

Floor plate

Roof plate

Incipient dorsal root
collateralization zone

Dorsal
funiculus

Dorsal root
bifurcation zone

Migrating and settling
dorsal horn neurons

Lateral cervical
nucleus?

Dorsal
root

Dorsal
neuroepithelium

Sulcus
limitans

Intermediate
interneurons

Intermediate
neuroepithelium

Sojourn zone
of premigratory
neurons destined
for the spinal gray

Dorsal
expansion

Central
canal

Ventral
expansion

Depleting ventral
interneuron source?

distal
limb

Ventral horn
interneurons
(settling)

Ventral horn
motoneurons
(segregating)

proximal
limb

axial

Depleted motoneuron
source?

Active source of
glial precursors?

Ventral
neuroepithelium

Lateral funiculus

Floor plate

Ventral commissure

Ventral funiculus

PLATE 10A

CR 17.5 mm
GW 7.8
M2155
Cervical
Cell body stain

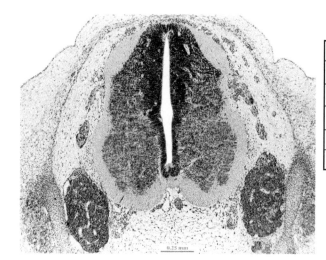

Areas (mm²)	
Central canal	.0510
Neuroepithelium	.2350
Roof plate	.0140
Floor plate	.0126
Gray matter	.6475
White matter	.3376

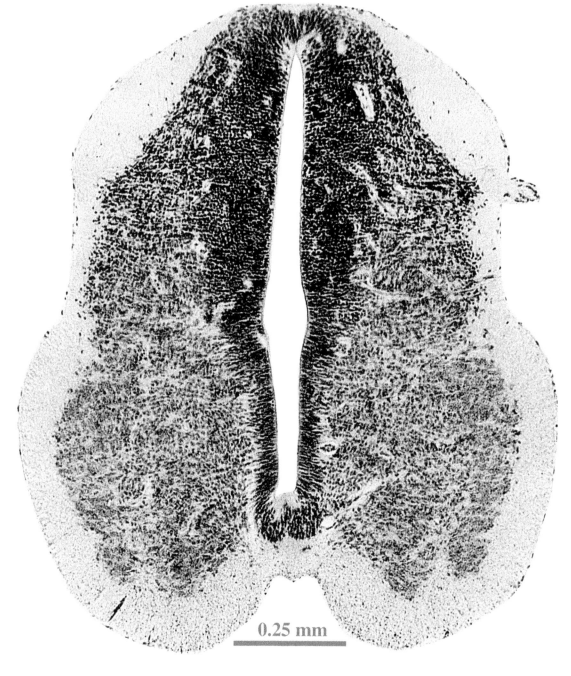

0.25 mm

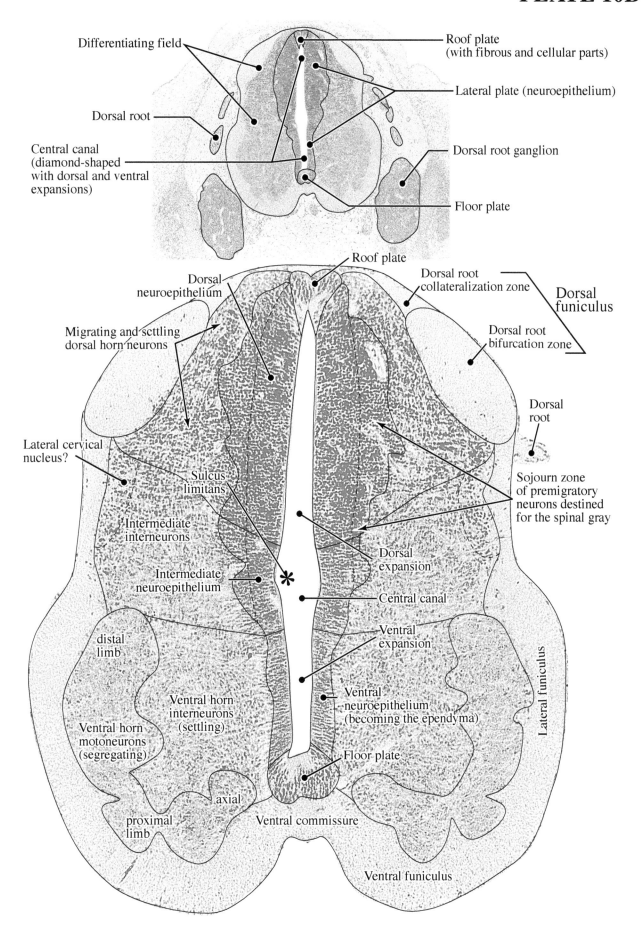

Differentiating field

Roof plate
(with fibrous and cellular parts)

Dorsal root

Lateral plate (neuroepithelium)

Central canal
(diamond-shaped
with dorsal and ventral
expansions)

Dorsal root ganglion

Floor plate

Roof plate

Dorsal
neuroepithelium

Dorsal root
collateralization zone

Dorsal
funiculus

Migrating and settling
dorsal horn neurons

Dorsal root
bifurcation zone

Dorsal
root

Lateral cervical
nucleus?

Sulcus
limitans

Intermediate
interneurons

Sojourn zone
of premigratory
neurons destined
for the spinal gray

Intermediate
neuroepithelium

Dorsal
expansion

*

Central canal

distal
limb

Ventral
expansion

Ventral horn
interneurons
(settling)

Ventral
neuroepithelium
(becoming the ependyma)

Lateral funiculus

Ventral horn
motoneurons
(segregating)

axial

Floor plate

proximal
limb

Ventral commissure

Ventral funiculus

PLATE 11A

CR 22.5 mm
GW 8.3
C7254
Cervical
Cell body stain

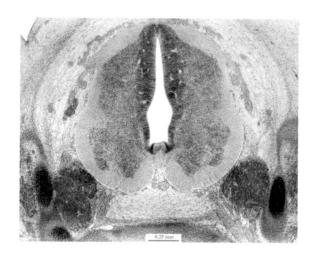

Areas (mm²)	
Central canal	.0887
Neuroepithelium	.2207
Roof plate	.0149
Floor plate	.0153
Gray matter	.6798
White matter	.4362

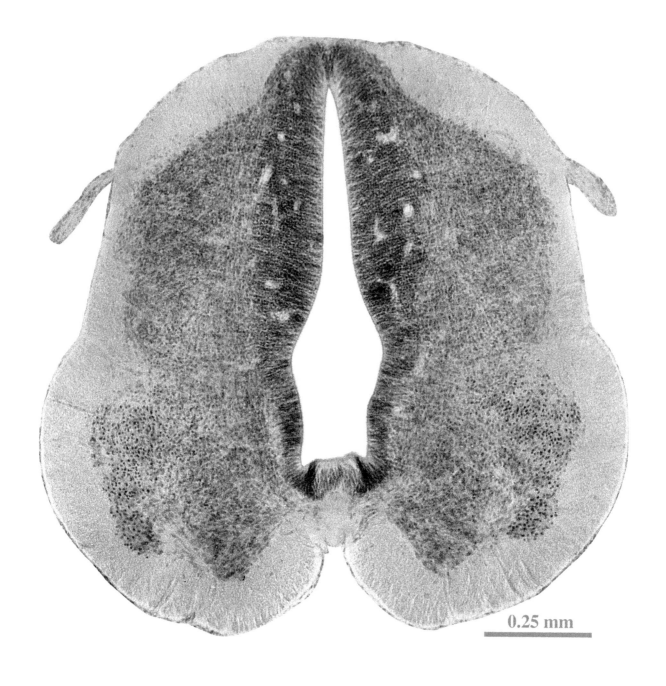

0.25 mm

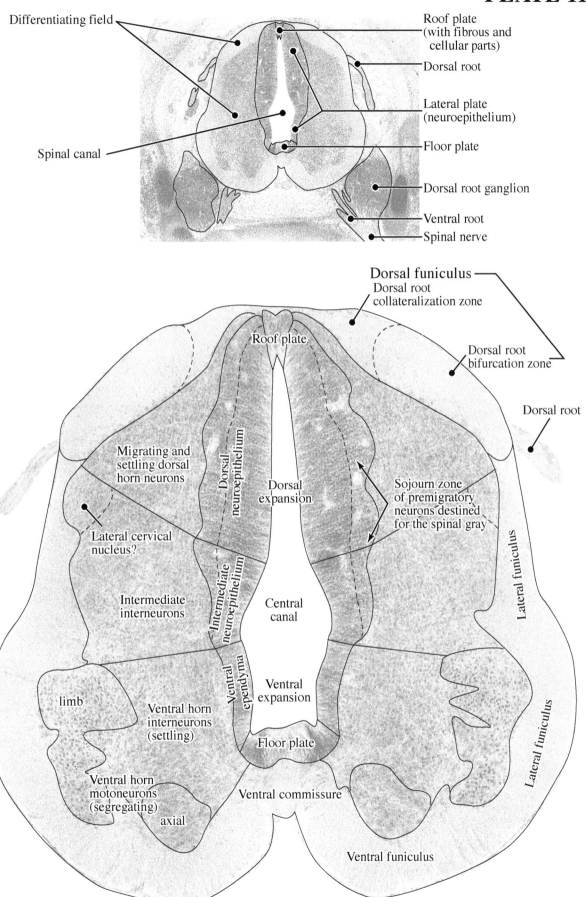

Differentiating field

Spinal canal

Roof plate
(with fibrous and
cellular parts)

Dorsal root

Lateral plate
(neuroepithelium)

Floor plate

Dorsal root ganglion

Ventral root

Spinal nerve

Dorsal funiculus

Dorsal root
collateralization zone

Dorsal root
bifurcation zone

Dorsal root

Roof plate

Migrating and
settling dorsal
horn neurons

Dorsal
neuroepithelium

Dorsal
expansion

Sojourn zone
of premigratory
neurons destined
for the spinal gray

Lateral cervical
nucleus?

Intermediate
interneurons

Intermediate
neuroepithelium

Central
canal

Lateral funiculus

limb

Ventral
ependyma

Ventral horn
interneurons
(settling)

Ventral
expansion

Floor plate

Lateral funiculus

Ventral horn
motoneurons
(segregating)

axial

Ventral commissure

Ventral funiculus

PLATE 12A

CR 32 mm
GW 9.6
C609
Cervical
Cell body stain

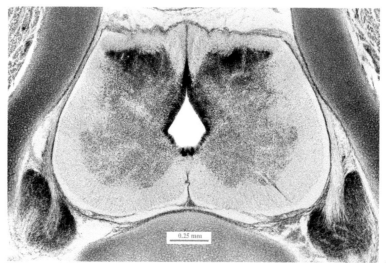

0.25 mm

Areas (mm²)

Central canal	.0439
Neuroepithelium	.0470
Roof plate	.0092
Floor plate	.0056
Gray matter	.8925
White matter	.6842

0.25 mm

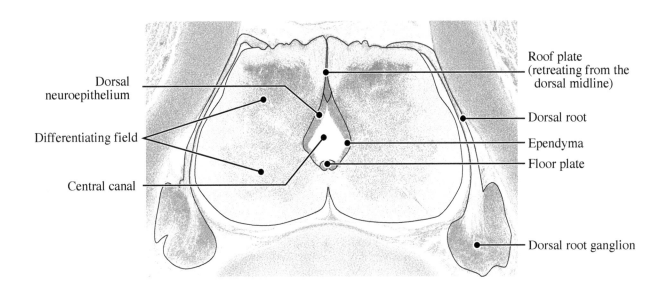

Dorsal
neuroepithelium

Differentiating field

Central canal

Roof plate
(retreating from the
dorsal midline)

Dorsal root

Ependyma

Floor plate

Dorsal root ganglion

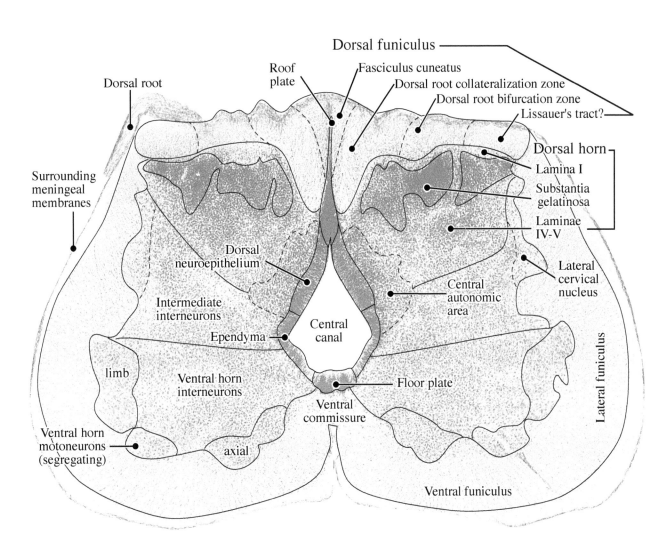

Dorsal funiculus

Roof
plate

Fasciculus cuneatus

Dorsal root collateralization zone

Dorsal root bifurcation zone

Lissauer's tract?

Dorsal root

Dorsal horn

Lamina I

Substantia
gelatinosa

Laminae
IV-V

Surrounding
meningeal
membranes

Dorsal
neuroepithelium

Central
autonomic
area

Lateral
cervical
nucleus

Intermediate
interneurons

Ependyma

Central
canal

limb

Ventral horn
interneurons

Floor plate

Lateral funiculus

Ventral
commissure

Ventral horn
motoneurons
(segregating)

axial

Ventral funiculus

PART III: M2050
CR 36 mm (GW 10)

Plate 13 is a survey of sections from seven levels of the spinal cord in M2050, a specimen in the Minot Collection-with a crown-rump length of 36 mm. and estimated to be at GW 10. All sections are shown at the same scale. The boxes enclosing each section list the level, ranging from upper cervical to coccygeal; the section number; and the total area of the section in square millimeters (mm^2). Note that the areal measurements are determined *after* fixation, while the crown-rump length is measured *before* fixation. The unfixed area of each section would probably be 40% to 60% larger. Thus the areal measurements are given only for comparison purposes between levels. Full-page normal-contrast photographs of each specimen are in Plates **14A–20A.** Low-contrast photographs with superimposed labels and outlines of structural details are in Plates **14B–20B.** This specimen is discussed in more detail in Chapter 5 of the companion reference book (Altman and Bayer, 2001).

By this time in spinal cord development, there are definite size differences between cervical, thoracic, lumbar, sacral, and coccygeal levels. Prior to GW 10, the spinal cord is progressively smaller from cervical to coccygeal levels. In M2050, there are still size differences that reflect the earlier developmental period. First, the cervical enlargement is about the same size as the upper cervical level. Second, the lumbar enlargement is 50% smaller than the cervical enlargement. But regional size differences in the adult spinal cord are beginning to show. For example, the middle thoracic level is 66% smaller than the cervical enlargement and 30% smaller than the lumbar enlargement.

The neuroepithelium is declining in this specimen. At all levels, there is still an active dorsal neuroepithelium. The basal edges of that neuroepithelium are indistinct as a multitude of cells, presumably young neurons destined for the dorsal horn and the central autonomic area, migrate away to settle. The basal "undulations" are prominent, which we postulate to be clumps of premigratory young neurons that sojourn in the neuroepithelium just after their generation. At low thoracic and lumbar levels, the upper part of the intermediate neuroepithelium still appears to be active. By the time one reaches the coccygeal level, even the ventral neuroepithelium appears to be active. Thus the very latest neurons were still being generated at the time of death in this specimen.

Concomitant with the decline of the neuroepithelium, an ependymal layer is appearing as low as lumbar levels. The ventral expansion of the spinal canal is absent except at sacral/coccygeal levels, and the dorsal expansion of the spinal canal is shortening. The central part of the canal is persisting, as it does in adults.

The gray matter is showing maturational characteristics of the adult spinal cord. In the ventral horn, the motoneurons are segregating into large clumps around the ventral and lateral borders. That is clearly seen in the cervical and lumbar enlargements. We postulate that the medial accumulations supply the axial muscles, while the lateral accumulations represent those that supply limb muscles. The thoracic ventral horn does not have lateral accumulations of motoneurons, but only medial ones supplying axial muscles. At cervical levels, the intermediate gray has a distinguishable lateral cervical nucleus, a migratory stream of large neurons that may be destined to settle in the lateral cervical nucleus, and an accumulation of large neurons near the central canal that may represent the central cervical nucleus. At thoracic levels, there is only a hint of some large neurons accumulating in Clarke's column (on the right side of the section in Plates 17A and 17B). Throughout all levels of the dorsal horn, there are distinctive clumps of densely packed small cells, presumably the neurons in laminae II and III, of the substantia gelatinosa. Indeed migratory streams of cells leaving the dorsal neuroepithelium appear to be heading for the substantia gelatinosa. The central autonomic area can be distinguished as a denser accumulation of small cells surrounding the central and dorsal parts of the spinal canal.

In the white matter, all components are growing larger. The most notable changes are occurring in the ventral commissure, ventral funiculus, and dorsal funiculus. The ventral commissure thickens beneath the floor plate, and moves upward as the ventral neuroepithelium and ventral spinal canal disappear. That creates a medial wall in the ventral funiculus that rapidly fills with axons. The dorsal funiculus is descending from the dorsal midline, following the retreating roof plate. Prior to GW 10, the roof plate is adjacent to the pial membrane in the dorsal midline. Now it extends upward in a sharp spike, and an accumulation of cells seem to form a string to the pia, defining the midline and possibly providing a structural separation between the dorsal funiculi on either side of the midline.

PLATE 13

CR 36 mm, GW 10, M2050

Plates 14A, 14B
Upper Cervical
Section 730
Total area:
2.8238 mm^2

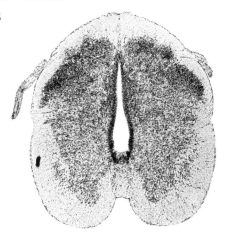

Plates 15A, 15B
Cervical Enlargement
Section 921
Total area:
2.8308 mm^2

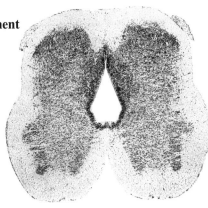
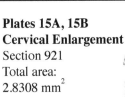

Plates 16A, 16B
Lower Cervical
Section 1113
Total area:
1.7018 mm^2

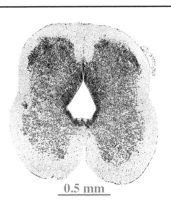

0.5 mm

Plates 17A, 17B
Middle Thoracic
Section 1593
Total area:
0.9715 mm^2

Plates 18A, 18B
Lower Thoracic
Section 1884
Total area:
1.1852 mm^2

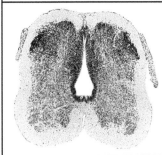

Plates 19A, 19B
Lumbar Enlargement
Section 2134
Total area:
1.3964 mm^2

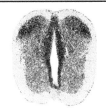

Plates 20A, 20B
(top sections)
Sacral
Section 2607
Total area:
0.6220 mm^2

Plates 20A, 20B
(bottom sections)
Sacral/Coccygeal
Section 2674
Total area:
0.3906 mm^2

0.5 mm

PLATE 14A

CR 36 mm
GW 10
M2050
Upper Cervical
Cell body stain

Areas (mm^2)	
Central canal	.0622
Neuroepithelium	.1072
Roof plate	.0190
Floor plate	.0209
Gray matter	1.7430
White matter	.8715

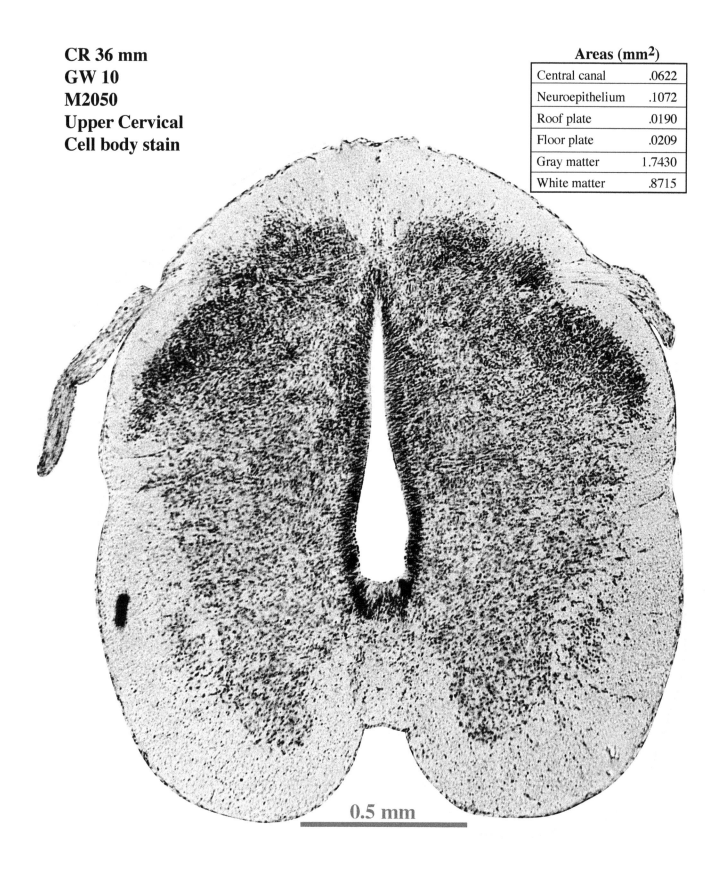

0.5 mm

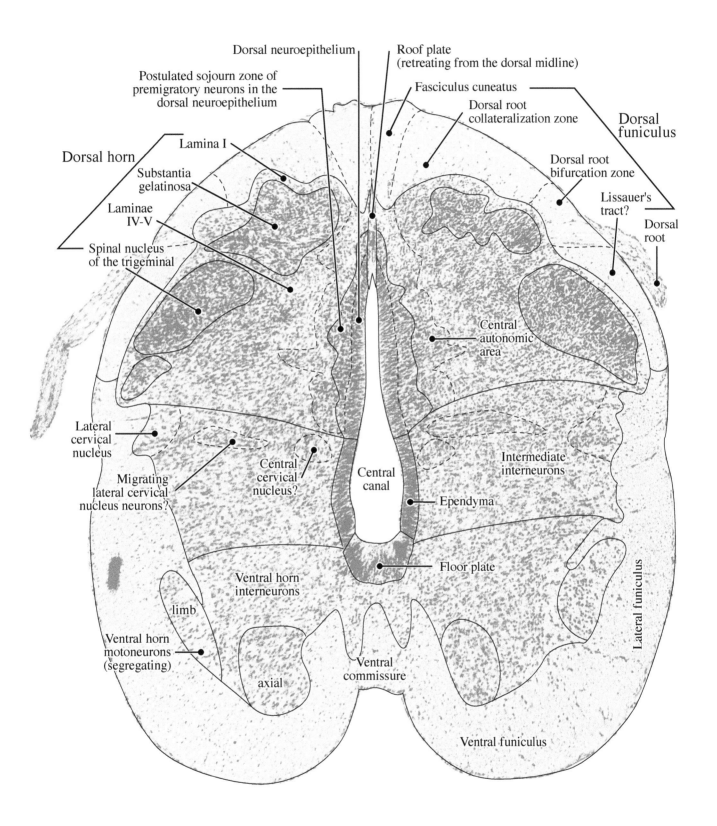

Dorsal neuroepithelium

Roof plate
(retreating from the dorsal midline)

Postulated sojourn zone of
premigratory neurons in the
dorsal neuroepithelium

Fasciculus cuneatus

Dorsal root
collateralization zone

Dorsal
funiculus

Dorsal horn

Lamina I

Substantia
gelatinosa

Dorsal root
bifurcation zone

Laminae
IV-V

Lissauer's
tract?

Spinal nucleus
of the trigeminal

Dorsal
root

Central
autonomic
area

Lateral
cervical
nucleus

Intermediate
interneurons

Central
cervical
nucleus?

Central
canal

Ependyma

Migrating
lateral cervical
nucleus neurons?

Ventral horn
interneurons

Floor plate

Lateral funiculus

limb

Ventral horn
motoneurons
(segregating)

axial

Ventral
commissure

Ventral funiculus

PLATE 15A

CR 36 mm
GW 10
M2050
Cervical Enlargement
Cell body stain

Areas (mm^2)	
Central canal	.0834
Neuroepithelium	.0966
Roof plate	.0221
Floor plate	.0161
Gray matter	1.5254
White matter	1.0872

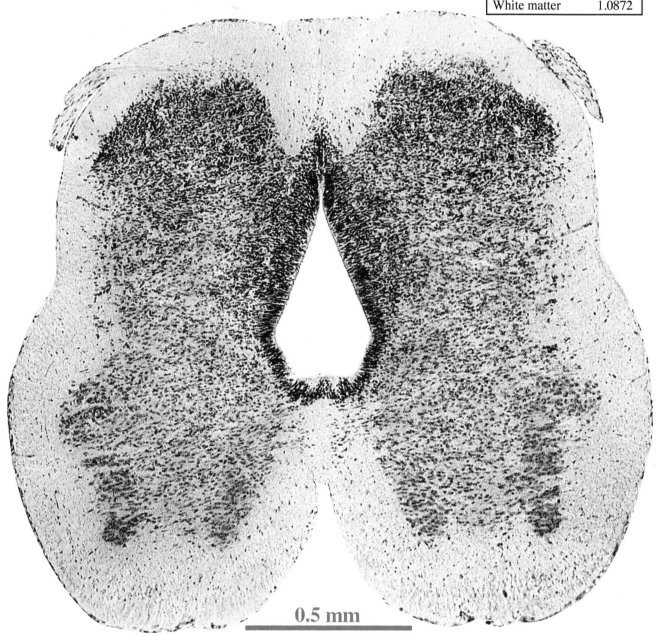

0.5 mm

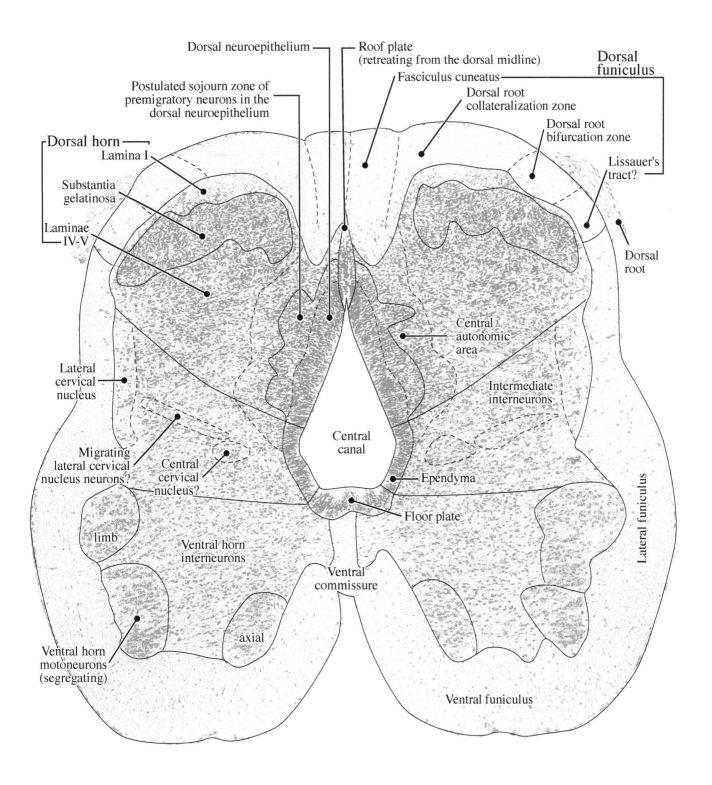

Dorsal neuroepithelium

Roof plate
(retreating from the dorsal midline)

Fasciculus cuneatus

Dorsal
funiculus

Postulated sojourn zone of
premigratory neurons in the
dorsal neuroepithelium

Dorsal root
collateralization zone

Dorsal root
bifurcation zone

Dorsal horn

Lissauer's
tract?

Lamina I

Substantia
gelatinosa

Laminae
IV-V

Dorsal
root

Central
autonomic
area

Lateral
cervical
nucleus

Intermediate
interneurons

Migrating
lateral cervical
nucleus neurons?

Central
cervical
nucleus?

Central
canal

Ependyma

Floor plate

Lateral funiculus

limb

Ventral horn
interneurons

Ventral
commissure

axial

Ventral horn
motoneurons
(segregating)

Ventral funiculus

PLATE 16A

CR 36 mm
GW 10
M2050
Lower Cervical
Cell body stain

Areas (mm²)	
Central canal	.0557
Neuroepithelium	.0874
Roof plate	.0124
Floor plate	.0156
Gray matter	.8918
White matter	.6389

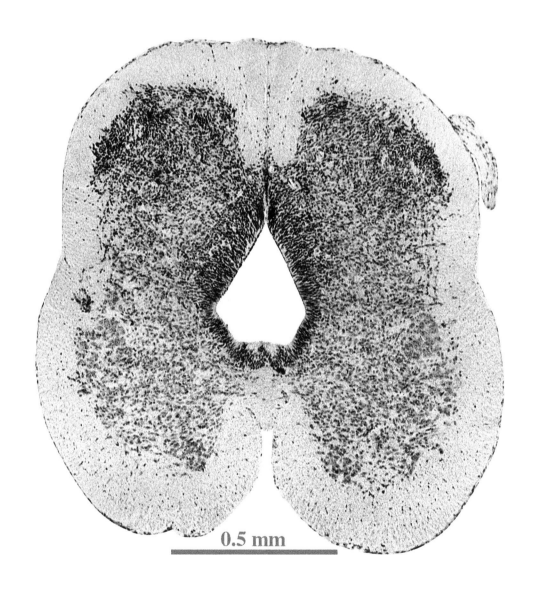

0.5 mm

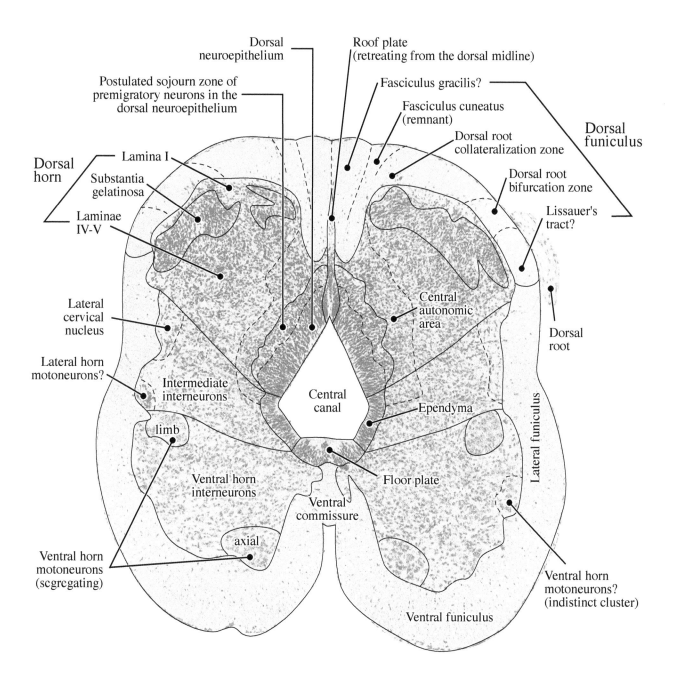

Dorsal
neuroepithelium

Roof plate
(retreating from the dorsal midline)

Postulated sojourn zone of
premigratory neurons in the
dorsal neuroepithelium

Fasciculus gracilis?

Fasciculus cuneatus
(remnant)

Dorsal root
collateralization zone

Dorsal
funiculus

Dorsal
horn

Lamina I

Substantia
gelatinosa

Dorsal root
bifurcation zone

Laminae
IV-V

Lissauer's
tract?

Lateral
cervical
nucleus

Central
autonomic
area

Dorsal
root

Lateral horn
motoneurons?

Intermediate
interneurons

Central
canal

Ependyma

Lateral funiculus

limb

Ventral horn
interneurons

Floor plate

Ventral
commissure

axial

Ventral horn
motoneurons
(scgregating)

Ventral horn
motoneurons?
(indistinct cluster)

Ventral funiculus

PLATE 17A

CR 36 mm
GW 10
M2050
Middle Thoracic
Cell body stain

Areas (mm^2)	
Central canal	.0238
Neuroepithelium	.0625
Roof plate	.0094
Floor plate	.0110
Gray matter	.4796
White matter	.3852

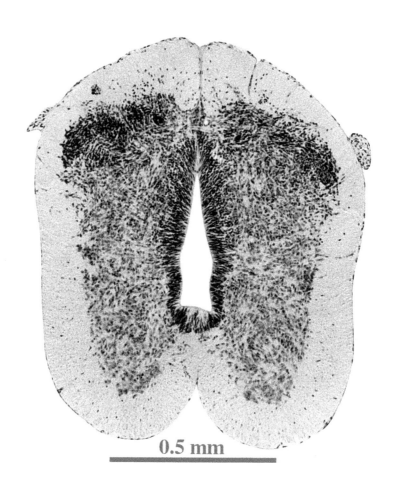

0.5 mm

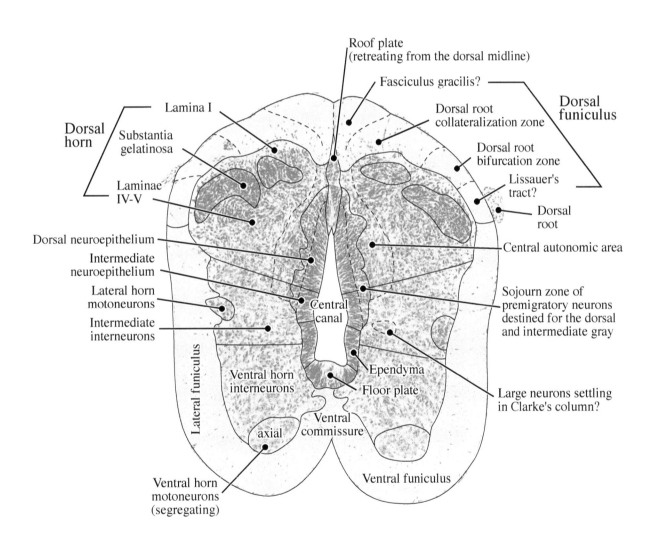

Roof plate
(retreating from the dorsal midline)

Fasciculus gracilis?

Dorsal
funiculus

Lamina I

Dorsal
horn

Substantia
gelatinosa

Dorsal root
collateralization zone

Dorsal root
bifurcation zone

Lissauer's
tract?

Laminae
IV-V

Dorsal
root

Dorsal neuroepithelium

Central autonomic area

Intermediate
neuroepithelium

Lateral horn
motoneurons

Sojourn zone of
premigratory neurons
destined for the dorsal
and intermediate gray

Intermediate
interneurons

Central
canal

Lateral funiculus

Ventral horn
interneurons

Ependyma

Large neurons settling
in Clarke's column?

Floor plate

axial

Ventral
commissure

Ventral horn
motoneurons
(segregating)

Ventral funiculus

PLATE 18A

CR 36 mm
GW 10
M2050
Lower Thoracic
Cell body stain

Areas (mm^2)	
Central canal	.0317
Neuroepithelium	.0857
Roof plate	.0097
Floor plate	.0132
Gray matter	.6133
White matter	.4317

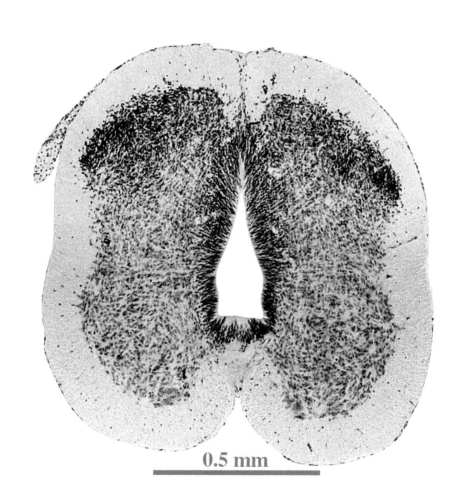

0.5 mm

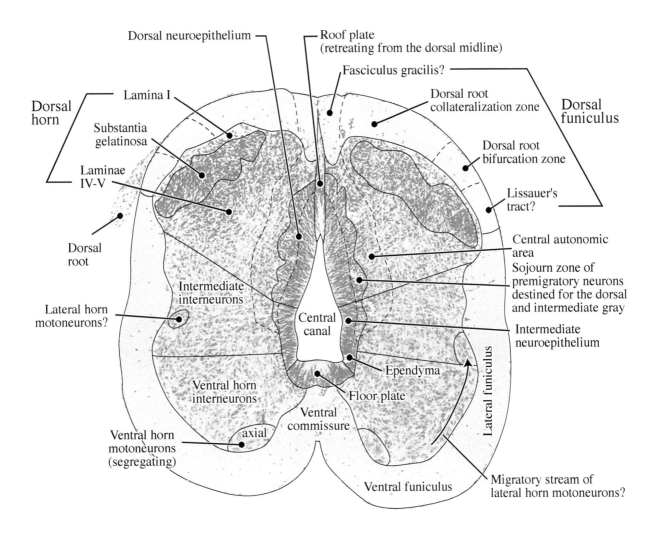

Dorsal neuroepithelium

Roof plate
(retreating from the dorsal midline)

Fasciculus gracilis?

Dorsal root
collateralization zone

Dorsal
funiculus

Dorsal
horn

Lamina I

Substantia
gelatinosa

Dorsal root
bifurcation zone

Laminae
IV-V

Lissauer's
tract?

Dorsal
root

Central autonomic
area

Sojourn zone of
premigratory neurons
destined for the dorsal
and intermediate gray

Lateral horn
motoneurons?

Intermediate
interneurons

Central
canal

Intermediate
neuroepithelium

Ependyma

Ventral horn
interneurons

Floor plate

Ventral
commissure

Lateral funiculus

axial

Ventral horn
motoneurons
(segregating)

Ventral funiculus

Migratory stream of
lateral horn motoneurons?

PLATE 19A

CR 36 mm
GW10
M2050
Lumbar Enlargement
Cell body stain

Areas (mm^2)	
Central canal	.0353
Neuroepithelium	.0957
Roof plate	.0238
Floor plate	.0113
Gray matter	.7637
White matter	.4666

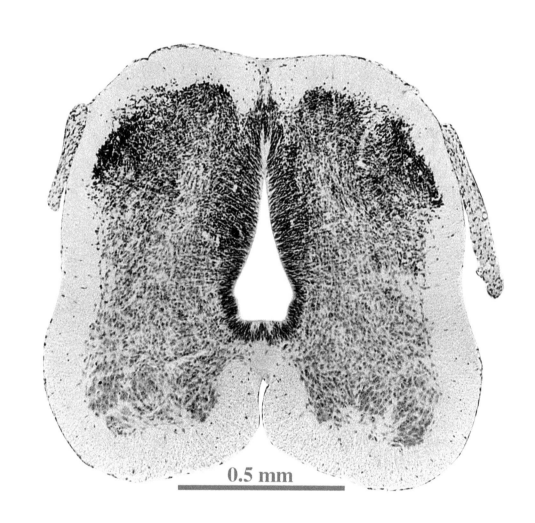

0.5 mm

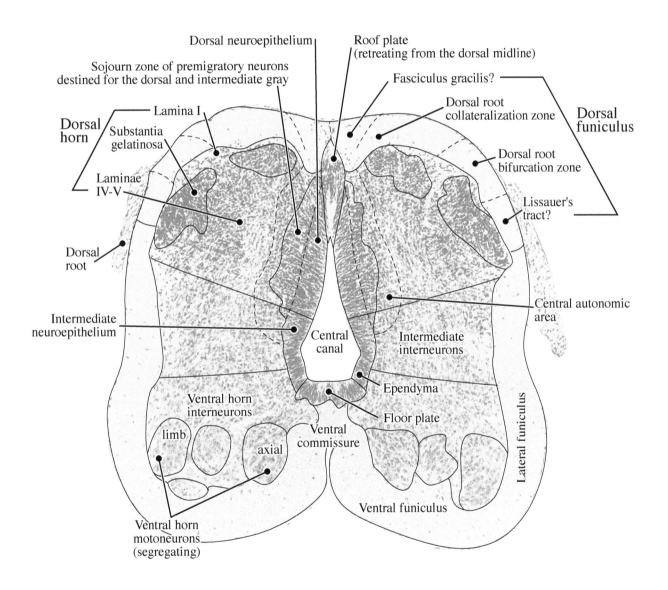

Dorsal neuroepithelium

Roof plate
(retreating from the dorsal midline)

Sojourn zone of premigratory neurons
destined for the dorsal and intermediate gray

Fasciculus gracilis?

Lamina I

Dorsal root
collateralization zone

Dorsal
horn

Substantia
gelatinosa

Dorsal
funiculus

Laminae
IV-V

Dorsal root
bifurcation zone

Lissauer's
tract?

Dorsal
root

Central autonomic
area

Intermediate
neuroepithelium

Intermediate
interneurons

Central
canal

Ependyma

Ventral horn
interneurons

limb

axial

Floor plate

Ventral
commissure

Lateral funiculus

Ventral funiculus

Ventral horn
motoneurons
(segregating)

PLATE 20A

CR 36 mm
GW 10
M2050
Sacral
Cell body stain

Areas (mm²)

Central canal	.0184
Neuroepithelium	.0647
Roof plate	.0135
Floor plate	.0066
Gray matter	.3275
White matter	.1913

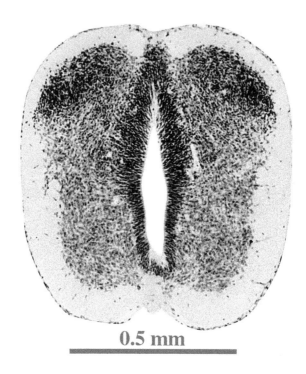

0.5 mm

Sacral/Coccygeal
Cell body stain

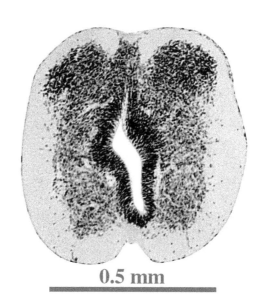

Areas (mm²)

Central canal	.0089
Neuroepithelium	.0386
Roof plate	.0052
Floor plate	.0029
Gray matter	.1962
White matter	.1389

0.5 mm

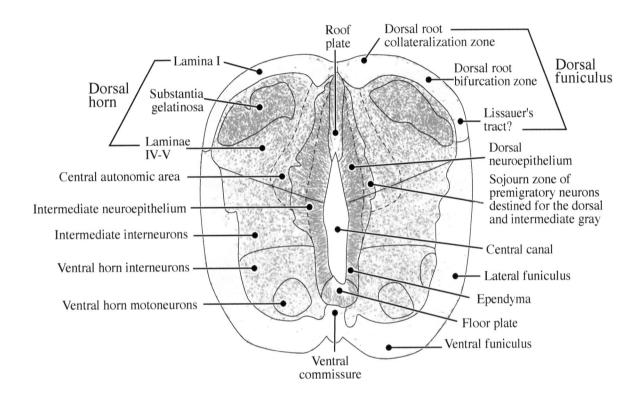

Roof plate
Dorsal root collateralization zone
Lamina I
Dorsal horn
Substantia gelatinosa
Dorsal funiculus
Dorsal root bifurcation zone
Lissauer's tract?
Laminae IV-V
Central autonomic area
Dorsal neuroepithelium
Intermediate neuroepithelium
Sojourn zone of premigratory neurons destined for the dorsal and intermediate gray
Intermediate interneurons
Ventral horn interneurons
Central canal
Ventral horn motoneurons
Lateral funiculus
Ependyma
Floor plate
Ventral funiculus
Ventral commissure

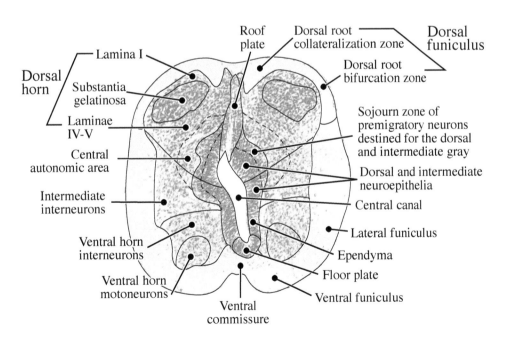

Roof plate
Dorsal root collateralization zone
Dorsal funiculus
Lamina I
Dorsal horn
Dorsal root bifurcation zone
Substantia gelatinosa
Laminae IV-V
Sojourn zone of premigratory neurons destined for the dorsal and intermediate gray
Central autonomic area
Dorsal and intermediate neuroepithelia
Intermediate interneurons
Central canal
Lateral funiculus
Ventral horn interneurons
Ependyma
Ventral horn motoneurons
Floor plate
Ventral funiculus
Ventral commissure

PART IV: Y380-62
CR 56 mm (GW 11.9)

GW 11.9 Plate 21 is a survey of sections from seven levels of the spinal cord in Y380-62, a specimen in the Yakovlev Collection with a crown-rump length of 56 mm. All sections are shown at the same scale. The boxes enclosing each section list the level ranging from upper cervical to coccygeal, the section number, and the total area of the section in square millimeters (mm^2). Note that the areal measurements are determined *after* fixation, while the crown-rump length is measured *before* fixation. The unfixed area of each section would probably be 40% to 60% larger. Thus the areal measurements are given only for comparison purposes between levels. Full-page normal-contrast photographs of each specimen are in Plates **22A–28A.** Low-contrast photographs with superimposed labels and outlines of structural details are in Plates **22B–28B.** This specimen is discussed in more detail in Chapter 5 of the companion reference book (Altman and Bayer, 2001).

The same size differences between levels seen at GW 10 are in this specimen. There are two size differences that reflect a developing rather than a mature spinal cord. First, the cervical enlargement is smaller than the upper cervical/lower medulla-level. Second, the lumbar enlargement is 11% smaller than the cervical enlargement, and the smallest cross-sectional area is at the sacral level. The lower thoracic level is 37% smaller than the cervical enlargement and 29% smaller than the lumbar enlargement, reflecting regional size differences characteristic of the adult spinal cord.

The neuroepithelium is no longer present in this specimen. At all levels, the central canal is shaped like a triangle, and is lined by a dense layer of cells, presumably the ependyma. However, proliferating cells interspersed among the ependyma may also be the source of glia. The lining of the central canal is labeled as ependyma in every section of this specimen.

The gray matter is continuing to show maturational changes that characterize the adult spinal cord. In the ventral horn, the motoneurons are segregating into more discrete motor columns than in the GW 10 specimen. The thoracic ventral horn has a large accumulation of motoneurons in the medial motor column. At cervical levels, the intermediate gray has a distinguishable lateral cervical nucleus, and an accumulation of large neurons near the central canal that may represent the central cervical nucleus. However, the migratory stream of large neurons that appeared to be moving into the lateral cervical nucleus (seen in the GW 10 specimen) cannot be distinguished. The dense groups of cells that appear to be migrating to the substantia gelatinosa, prominent at GW 10, are less distinct in this more mature specimen.

In the white matter, all components continue to grow larger. The dorsal funiculus continues to fill in with fibers as it descends from the dorsal midline, following the continually retreating roof plate.

CR 56mm, GW 11.9, Y380-62

Plates 22A, 22B
Lower Medulla/
Upper Cervical
Section 361
Total Area:
3.2032 mm^2

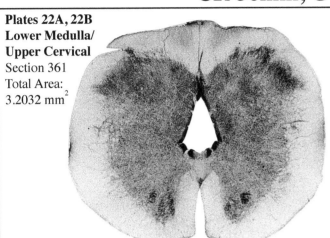

Plates 23A, 23B
Cervical Enlargement
Section 521
Total Area:
1.9640 mm^2

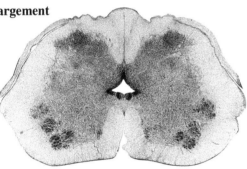

Plates 24A, 24B
Upper Thoracic
Section 671
Total Area:
1.4213 mm^2

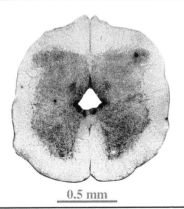

0.5 mm

Plates 25A, 25B
Lower Thoracic
Section 881
Total Area:
1.2365 mm^2

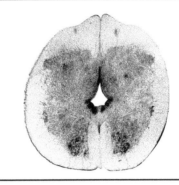

Plates 26A, 26B
Upper Lumbar
Section 911
Total Area:
1.4099 mm^2

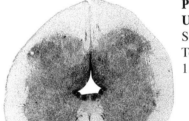

Plates 27A, 27B
Lumbar
Enlargement
Section 1066
Total Area:
1.7510 mm^2

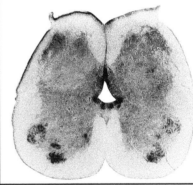

Plates 28A, 28B
Sacral
Section 1211
Total Area:
0.6813 mm^2

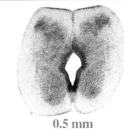

0.5 mm

PLATE 22A

CR 56 mm
GW 11.9
Y380-62
Lower Medulla/Upper Cervical
Cell body stain

Areas (mm²)	
Central canal	.0728
Neuroepithelium	.0395
Roof plate	.0199
Floor plate	.0139
Gray matter	1.7140
White matter	1.3431

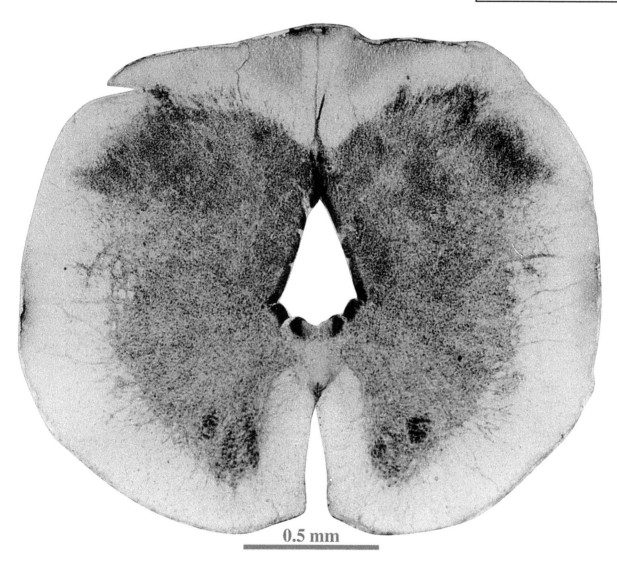

0.5 mm

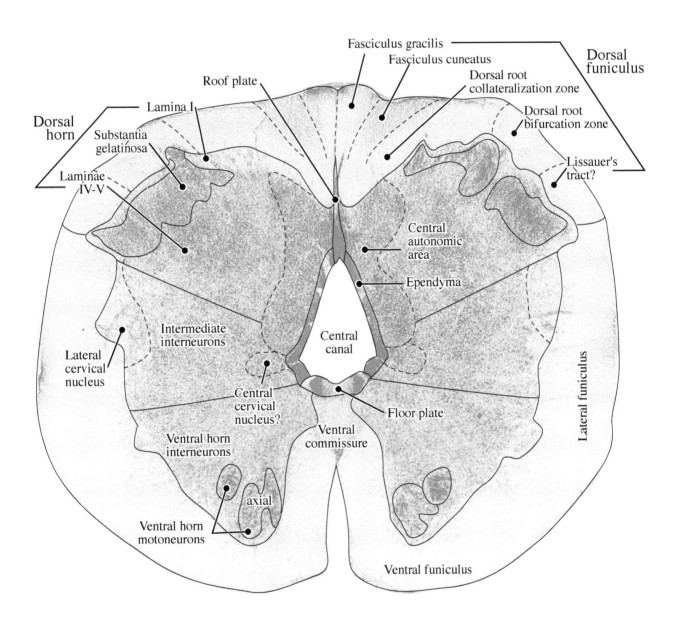

Fasciculus gracilis

Fasciculus cuneatus

Dorsal root collateralization zone

Dorsal funiculus

Roof plate

Dorsal root bifurcation zone

Lamina I

Dorsal horn

Substantia gelatinosa

Laminae IV-V

Lissauer's tract?

Central autonomic area

Ependyma

Intermediate interneurons

Central canal

Lateral cervical nucleus

Central cervical nucleus?

Floor plate

Lateral funiculus

Ventral horn interneurons

Ventral commissure

axial

Ventral horn motoneurons

Ventral funiculus

PLATE 23A

CR 56 mm
GW 11.9
Y380-62
Cervical Enlargement
Cell body stain

Areas (mm^2)	
Central canal	.0105
Neuroepithelium	.0135
Roof plate	.0133
Floor plate	.0086
Gray matter	1.0320
White matter	.8861

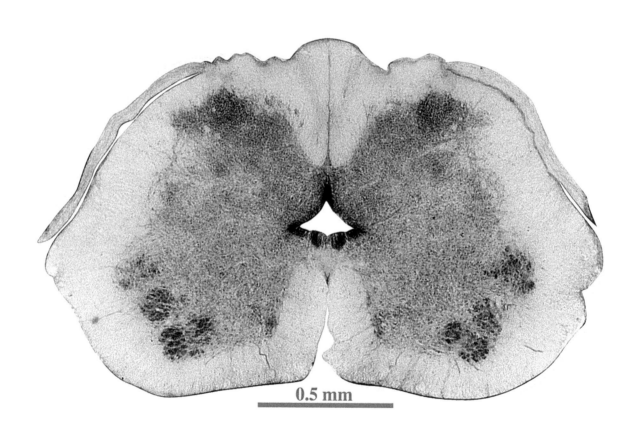

0.5 mm

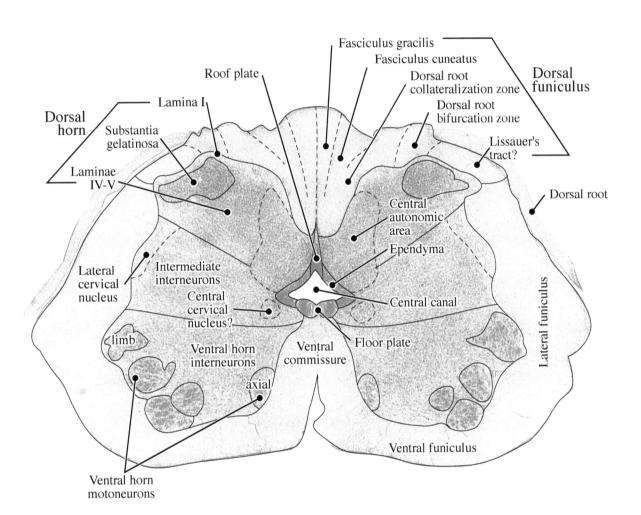

Fasciculus gracilis

Fasciculus cuneatus

Roof plate

Dorsal root
collateralization zone

Dorsal
funiculus

Dorsal root
bifurcation zone

Lamina I

Dorsal
horn

Substantia
gelatinosa

Lissauer's
tract?

Laminae
IV-V

Dorsal root

Central
autonomic
area

Ependyma

Lateral
cervical
nucleus

Intermediate
interneurons

Central
cervical
nucleus?

Central canal

Lateral funiculus

limb.

Ventral horn
interneurons

Floor plate

axial

Ventral
commissure

Ventral horn
motoneurons

Ventral funiculus

PLATE 24A

CR 56 mm
GW 11.9
Y380-62
Upper Thoracic
Cell body stain

Areas (mm²)	
Central canal	.0206
Neuroepithelium	.0198
Roof plate	.0143
Floor plate	.0083
Gray matter	.6812
White matter	.6771

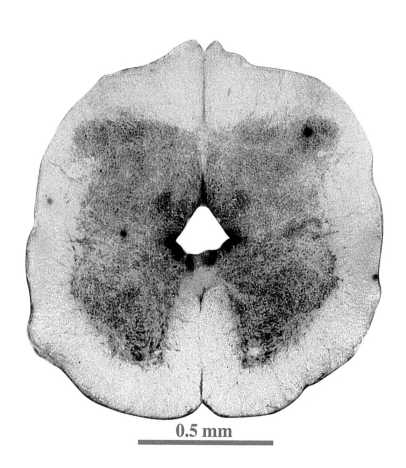

0.5 mm

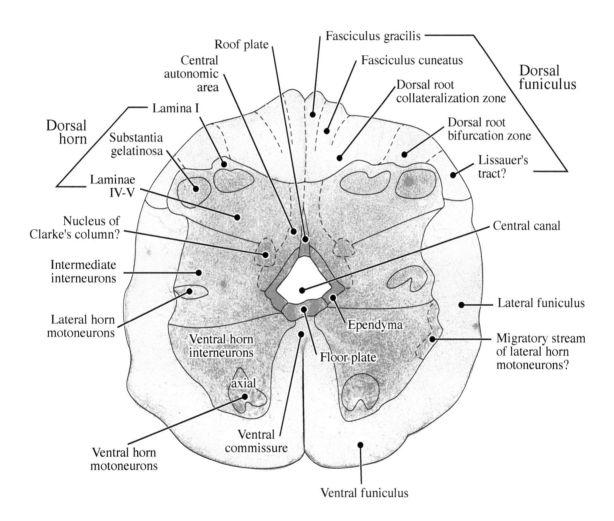

Roof plate
Central
autonomic
area
Fasciculus gracilis
Fasciculus cuneatus
Dorsal
funiculus
Lamina I
Dorsal root
collateralization zone
Dorsal
horn
Substantia
gelatinosa
Dorsal root
bifurcation zone
Laminae
IV-V
Lissauer's
tract?
Nucleus of
Clarke's column?
Central canal
Intermediate
interneurons
Lateral horn
motoneurons
Lateral funiculus
Ventral horn
interneurons
Ependyma
Migratory stream
of lateral horn
motoneurons?
axial
Floor plate
Ventral horn
motoneurons
Ventral
commissure
Ventral funiculus

PLATE 25A

CR 56 mm
GW 11.9
Y380-62
Lower Thoracic
Cell body stain

Areas (mm^2)	
Central canal	.0114
Neuroepithelium	.0207
Roof plate	.0100
Floor plate	.0082
Gray matter	.6527
White matter	.5336

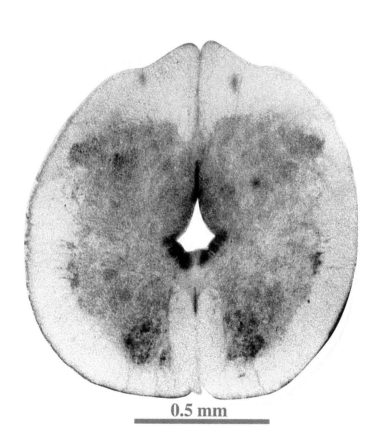

0.5 mm

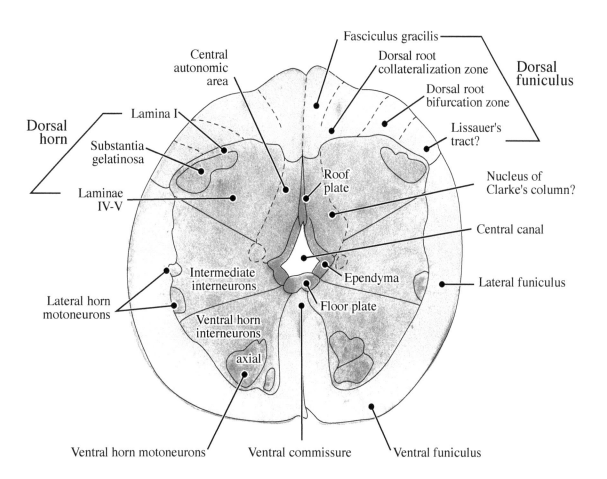

Fasciculus gracilis

Central
autonomic
area

Dorsal root
collateralization zone

Dorsal
funiculus

Dorsal root
bifurcation zone

Lamina I

Lissauer's
tract?

Dorsal
horn

Substantia
gelatinosa

Roof
plate

Nucleus of
Clarke's column?

Laminae
IV-V

Central canal

Ependyma

Lateral funiculus

Intermediate
interneurons

Lateral horn
motoneurons

Floor plate

Ventral horn
interneurons

axial

Ventral horn motoneurons

Ventral commissure

Ventral funiculus

PLATE 26A

CR 56 mm
GW 11.9
Y380-62
Upper Lumbar
Cell body stain

Areas (mm^2)	
Central canal	.0129
Neuroepithelium	.0232
Roof plate	.0143
Floor plate	.0091
Gray matter	.7558
White matter	.5992

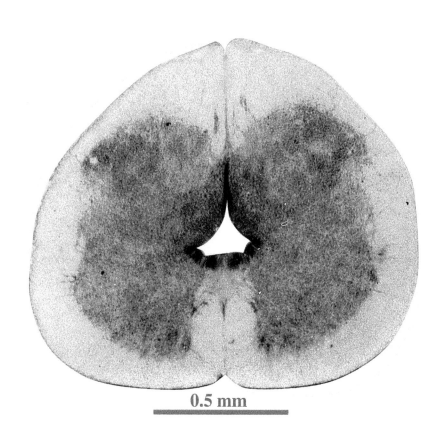

0.5 mm

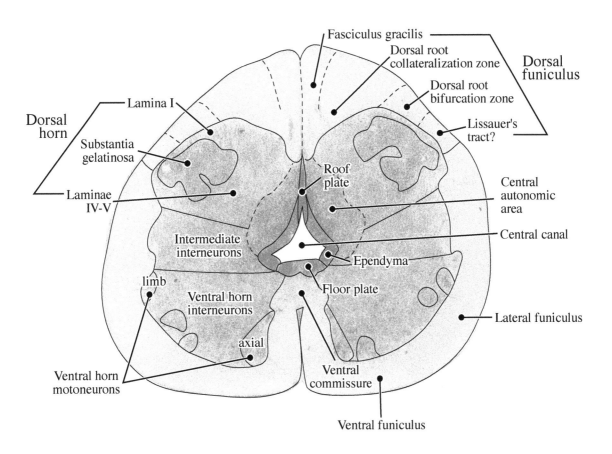

Fasciculus gracilis

Dorsal root
collateralization zone

Dorsal root
bifurcation zone

Dorsal
funiculus

Lamina I

Dorsal
horn

Substantia
gelatinosa

Lissauer's
tract?

Laminae
IV-V

Roof
plate

Central
autonomic
area

Central canal

Intermediate
interneurons

Ependyma

limb

Ventral horn
interneurons

Floor plate

Lateral funiculus

axial

Ventral horn
motoneurons

Ventral
commissure

Ventral funiculus

PLATE 27A

CR 56 mm
GW 11.9
Y380-62
Lumbar Enlargement
Cell body stain

Areas (mm^2)	
Central canal	.0128
Neuroepithelium	.0183
Roof plate	.0128
Floor plate	.0081
Gray matter	.9752
White matter	.7221

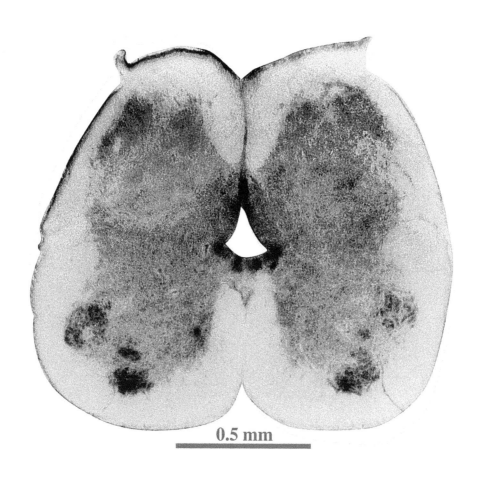

0.5 mm

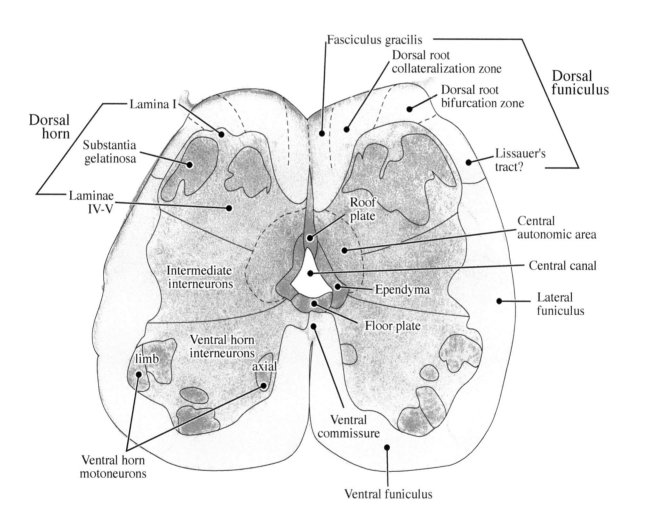

Fasciculus gracilis

Dorsal root
collateralization zone

Dorsal root
bifurcation zone

Dorsal
funiculus

Lamina I

Dorsal
horn

Substantia
gelatinosa

Lissauer's
tract?

Laminae
IV-V

Roof
plate

Central
autonomic area

Central canal

Intermediate
interneurons

Ependyma

Lateral
funiculus

Floor plate

Ventral horn
interneurons

limb

axial

Ventral horn
motoneurons

Ventral
commissure

Ventral funiculus

PLATE 28A

CR 56 mm
GW 11.9
Y380-62
Sacral
Cell body stain

Areas (mm²)	
Central canal	.0144
Neuroepithelium	.0311
Roof plate	.0201
Floor plate	.0058
Gray matter	.3385
White matter	.2754

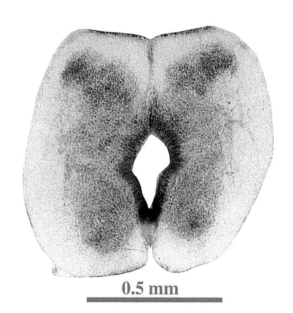

0.5 mm

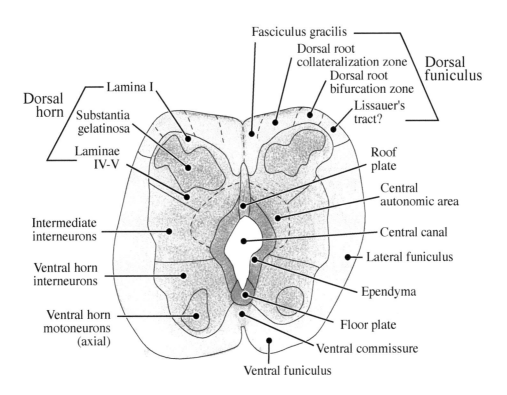

Fasciculus gracilis

Dorsal root
collateralization zone

Dorsal
funiculus

Dorsal root
bifurcation zone

Lamina I

Dorsal
horn

Lissauer's
tract?

Substantia
gelatinosa

Roof
plate

Laminae
IV-V

Central
autonomic area

Central canal

Intermediate
interneurons

Lateral funiculus

Ventral horn
interneurons

Ependyma

Ventral horn
motoneurons
(axial)

Floor plate

Ventral commissure

Ventral funiculus

PART V: Y68-65
CR 108 mm (GW 14)

Plate 29 is a survey of sections from Y68-65, a specimen in the Yakovlev Collection with a crown-rump length of 108 mm (see Chapters 5 and 6 of Altman and Bayer, 2001). This specimen is in the early second trimester and is included here because of the 3-D reconstructions presented later. All sections are shown at the same scale. The boxes enclosing each section list the level from upper cervical to sacral/coccygeal, the section number from the set of slides containing all the sections of that specimen, and the total area (post-fixation) of the section in square millimeters (mm²). Full-page normal-contrast photographs of each specimen are in **Plates 30A–37A**. Low-contrast photographs with superimposed labels and outlines of section details are on the facing pages in **Plates 30B–37B**.

The outer surface of this specimen has sharp spike-like projections, and the white matter is thin around the gray matter, interpreted to be a shrinkage artifact of histological processing. Cells in the gray matter are not distinct, except for the prominent clumps of large motoneurons in the ventral horn, possibly due to inadequate penetration of fixative. In spite of these histological artifacts, this specimen is illustrated because it is the only one in the Yakovlev Collection where sections of the spinal cord are consecutively numbered from cervical to coccygeal levels.

Most of the size differences between levels at GW 14 are similar to those in the adult spinal cord. The smallest cross-sectional area is in the middle thoracic level, which is 27% smaller than the sacral level. That the cervical enlargement level is smaller than both the upper cervical and lumbar enlargement levels reflects variability unique to this specimen. Within the gray matter, the most obvious sign of ongoing maturation is the prominent columnar arrangement of motoneurons in the ventral horn. Not only are these columns larger than at GW 10.5, but there are more of them. Within the white matter, the dorsal funiculus is deepening further in the dorsal midline, accompanying the retreating roof plate.

CR 108 mm, GW 14, Y68-65

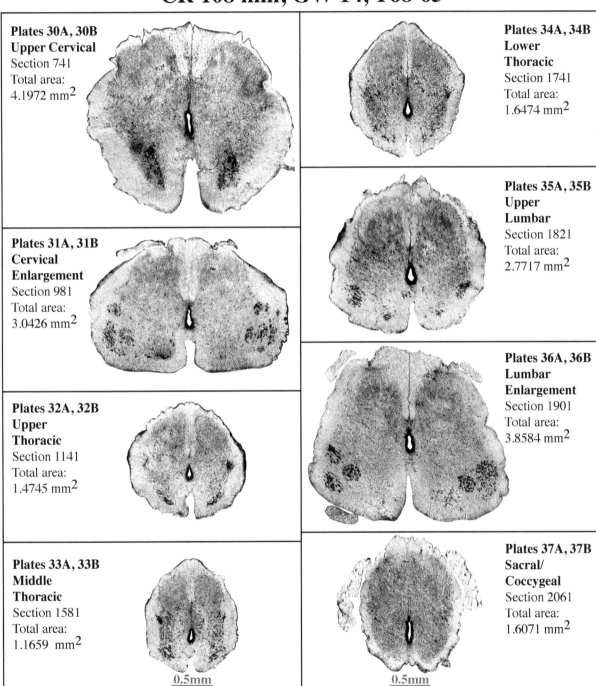

Plates 30A, 30B
Upper Cervical
Section 741
Total area:
4.1972 mm^2

Plates 31A, 31B
Cervical
Enlargement
Section 981
Total area:
3.0426 mm^2

Plates 32A, 32B
Upper
Thoracic
Section 1141
Total area:
1.4745 mm^2

Plates 33A, 33B
Middle
Thoracic
Section 1581
Total area:
1.1659 mm^2

0.5mm

Plates 34A, 34B
Lower
Thoracic
Section 1741
Total area:
1.6474 mm^2

Plates 35A, 35B
Upper
Lumbar
Section 1821
Total area:
2.7717 mm^2

Plates 36A, 36B
Lumbar
Enlargement
Section 1901
Total area:
3.8584 mm^2

Plates 37A, 37B
Sacral/
Coccygeal
Section 2061
Total area:
1.6071 mm^2

0.5mm

PLATE 30A

CR 108 mm
GW 14
Y68-65
Upper Cervical
Cell body stain

Areas (mm²)	
Central canal	.0053
Neuroepithelium	.0182
Roof plate	.0290
Floor plate	.0029
Gray matter	2.5381
White matter	1.6037

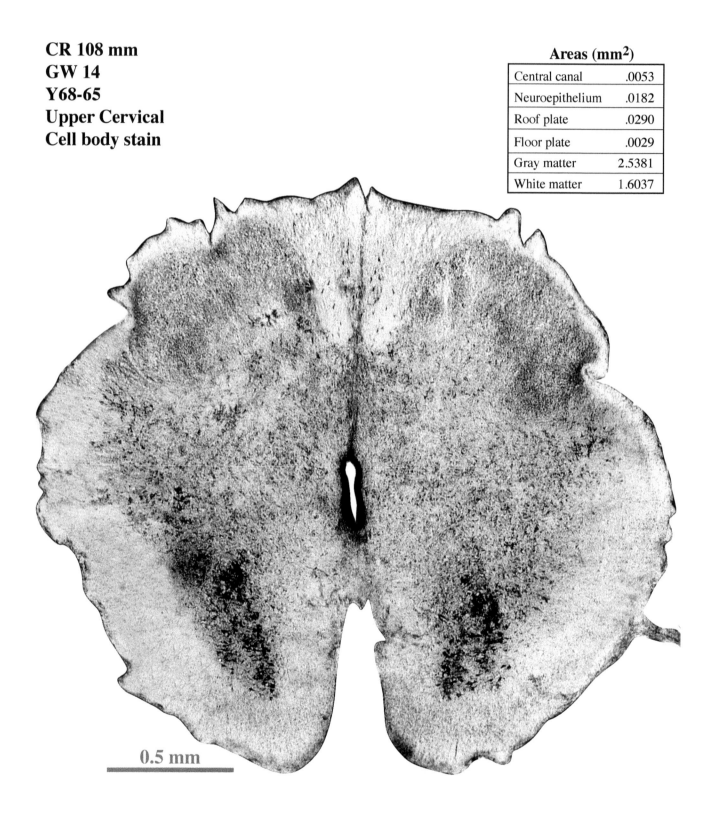

0.5 mm

PLATE 30B

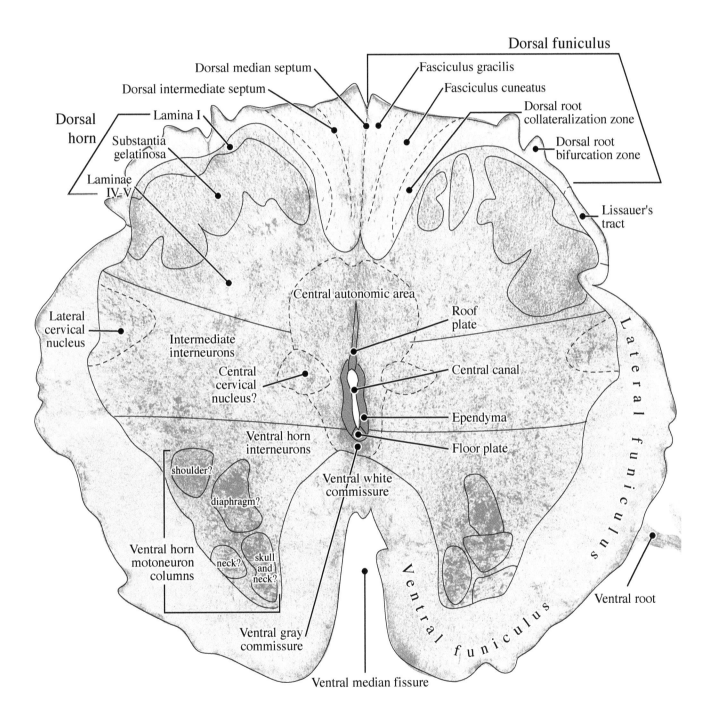

Dorsal funiculus

Dorsal median septum

Dorsal intermediate septum

Fasciculus gracilis

Fasciculus cuneatus

Dorsal horn

Lamina I

Substantia gelatinosa

Laminae IV–V

Dorsal root collateralization zone

Dorsal root bifurcation zone

Lissauer's tract

Central autonomic area

Roof plate

Lateral cervical nucleus

Intermediate interneurons

Central cervical nucleus?

Central canal

Ependyma

Ventral horn interneurons

Floor plate

shoulder?

diaphragm?

Ventral white commissure

neck?

skull and neck?

Ventral horn motoneuron columns

Ventral gray commissure

Lateral funiculus

Ventral funiculus

Ventral root

Ventral median fissure

PLATE 31A

CR 108 mm
GW 14
Y68-65
Cervical Enlargement
Cell body stain

Areas (mm^2)	
Central canal	.0056
Neuroepithelium	.0127
Roof plate	.0192
Floor plate	.0035
Gray matter	1.9505
White matter	1.0511

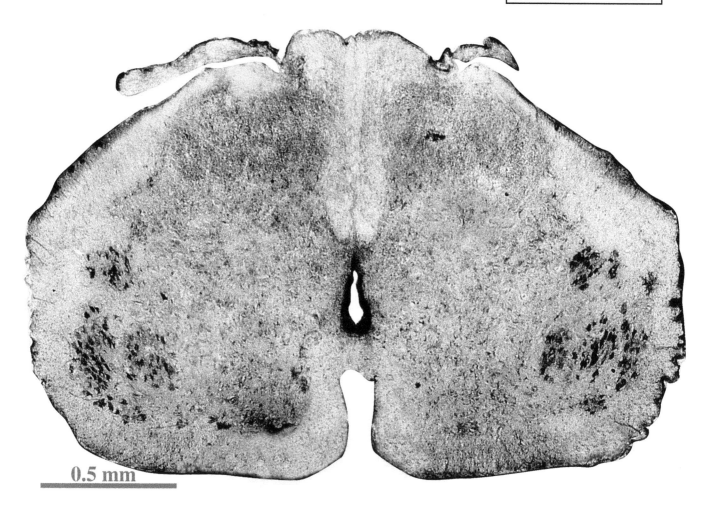

0.5 mm

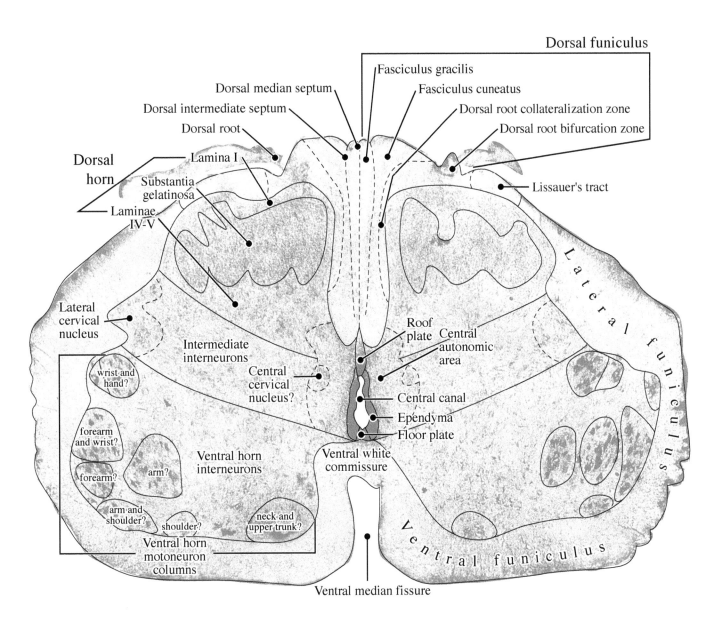

Dorsal funiculus

Fasciculus gracilis

Dorsal median septum

Fasciculus cuneatus

Dorsal intermediate septum

Dorsal root collateralization zone

Dorsal root

Dorsal root bifurcation zone

Dorsal horn

Lamina I

Lissauer's tract

Substantia gelatinosa

Laminae IV-V

Lateral cervical nucleus

Intermediate interneurons

Roof plate

Central autonomic area

Central cervical nucleus?

Central canal

Ependyma

Floor plate

wrist and hand?

forearm and wrist?

arm?

forearm?

Ventral horn interneurons

Ventral white commissure

arm and shoulder?

shoulder?

neck and upper trunk?

Ventral horn motoneuron columns

Lateral funiculus

Ventral funiculus

Ventral median fissure

PLATE 32A

CR 108 mm
GW 14
Y68-65
Upper Thoracic
Cell body stain

Areas (mm²)	
Central canal	.0021
Neuroepithelium	.0063
Roof plate	.0135
Floor plate	.0040
Gray matter	.8242
White matter	.6243

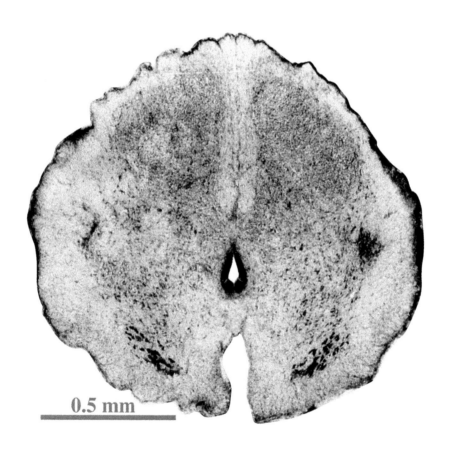

0.5 mm

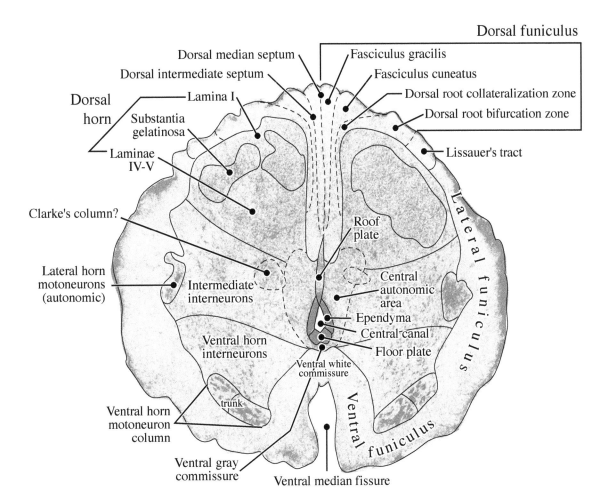

Dorsal funiculus

Dorsal median septum

Dorsal intermediate septum

Fasciculus gracilis

Fasciculus cuneatus

Dorsal horn

Lamina I

Dorsal root collateralization zone

Substantia gelatinosa

Dorsal root bifurcation zone

Laminae IV-V

Lissauer's tract

Clarke's column?

Roof plate

Lateral horn motoneurons (autonomic)

Central autonomic area

Intermediate interneurons

Ependyma

Central canal

Ventral horn interneurons

Floor plate

Ventral white commissure

trunk

Ventral horn motoneuron column

Ventral gray commissure

Ventral median fissure

Lateral funiculus

Ventral funiculus

PLATE 33A

CR 108 mm
GW 14
Y68-65
Middle Thoracic
Cell body stain

Areas (mm²)	
Central canal	.0018
Neuroepithelium	.0084
Roof plate	.0152
Floor plate	.0035
Gray matter	.6913
White matter	.4458

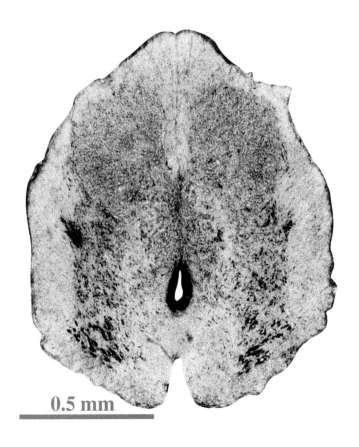

0.5 mm

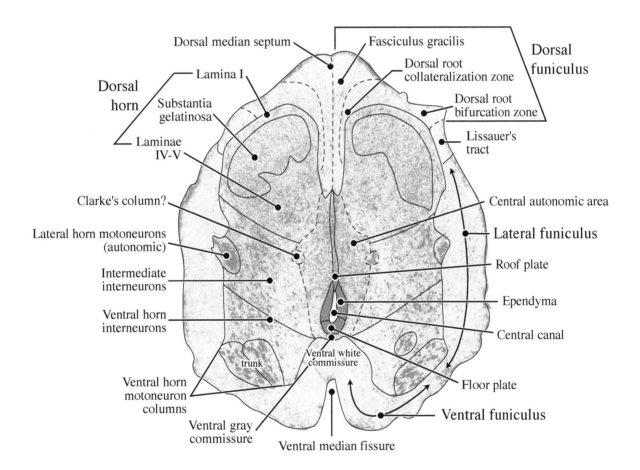

Dorsal median septum

Fasciculus gracilis

Dorsal root collateralization zone

Dorsal funiculus

Lamina I

Dorsal horn

Substantia gelatinosa

Dorsal root bifurcation zone

Laminae IV-V

Lissauer's tract

Clarke's column?

Central autonomic area

Lateral horn motoneurons (autonomic)

Lateral funiculus

Intermediate interneurons

Roof plate

Ependyma

Ventral horn interneurons

Central canal

trunk

Ventral white commissure

Floor plate

Ventral horn motoneuron columns

Ventral funiculus

Ventral gray commissure

Ventral median fissure

PLATE 34A

CR 108 mm
GW 14
Y68-65
Lower Thoracic
Cell body stain

Areas (mm^2)	
Central canal	.0039
Neuroepithelium	.0128
Roof plate	.0171
Floor plate	.0019
Gray matter	1.0249
White matter	.5868

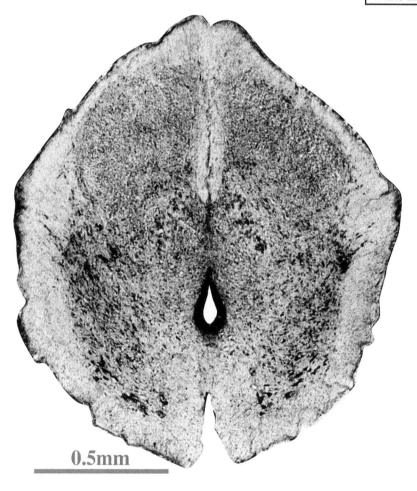

0.5mm

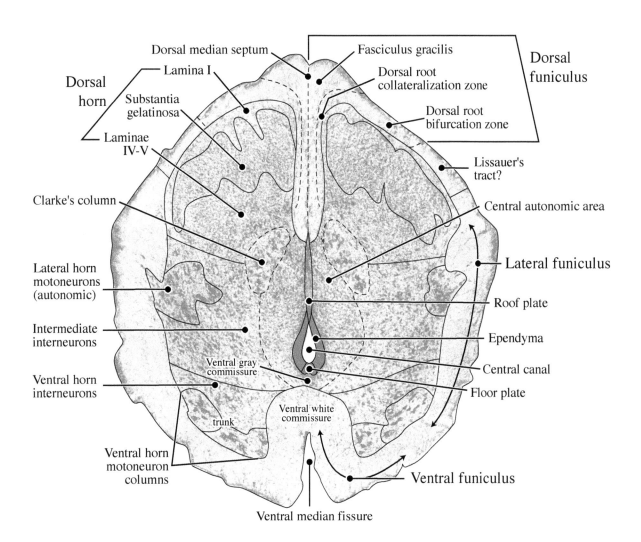

Dorsal median septum

Fasciculus gracilis

Dorsal
funiculus

Dorsal
horn

Lamina I

Substantia
gelatinosa

Dorsal root
collateralization zone

Dorsal root
bifurcation zone

Laminae
IV-V

Lissauer's
tract?

Clarke's column

Central autonomic area

Lateral horn
motoneurons
(autonomic)

Lateral funiculus

Roof plate

Intermediate
interneurons

Ependyma

Ventral horn
interneurons

Ventral gray
commissure

Central canal

Floor plate

trunk

Ventral white
commissure

Ventral horn
motoneuron
columns

Ventral funiculus

Ventral median fissure

PLATE 35A

CR 108 mm
GW 14
Y68-65
Upper Lumbar
Cell body stain

Areas (mm²)	
Central canal	.0056
Neuroepithelium	.0140
Roof plate	.0315
Floor plate	.0076
Gray matter	1.6817
White matter	1.0312

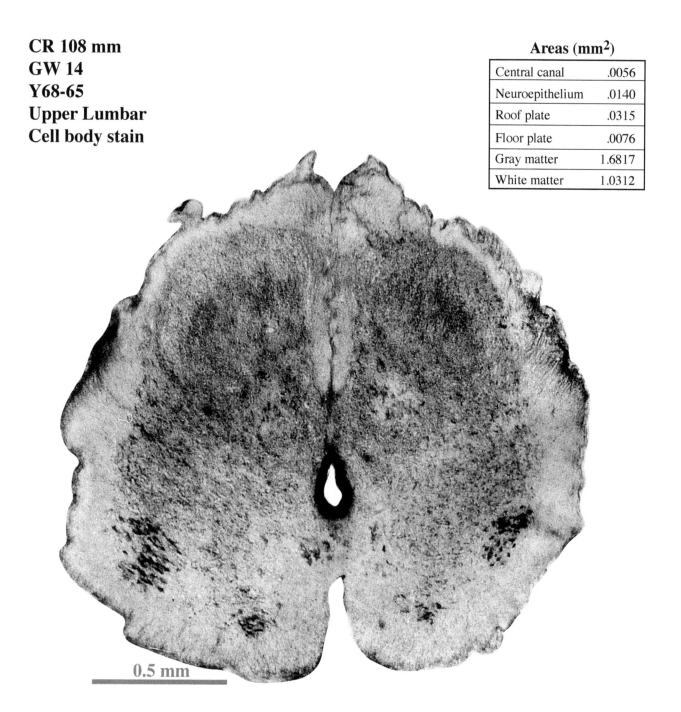

0.5 mm

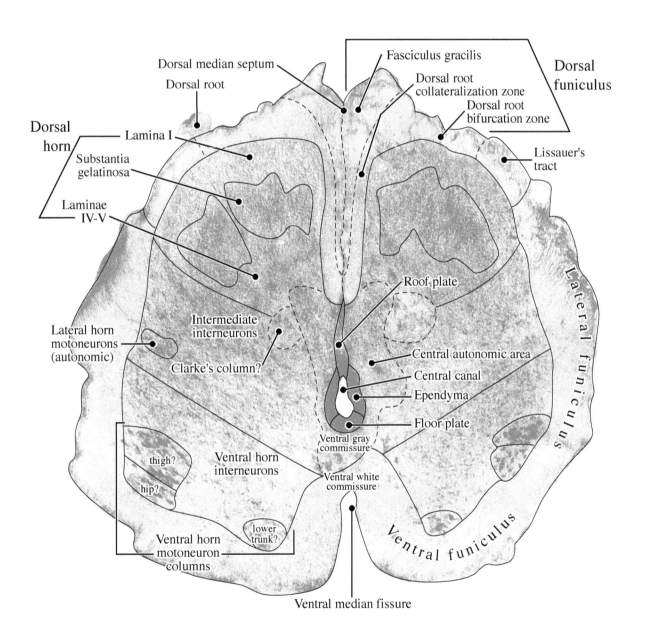

Dorsal median septum

Dorsal root

Fasciculus gracilis

Dorsal root
collateralization zone

Dorsal root
bifurcation zone

Dorsal
funiculus

Dorsal
horn

Lamina I

Substantia
gelatinosa

Laminae
IV-V

Lissauer's
tract

Lateral horn
motoneurons
(autonomic)

Intermediate
interneurons

Clarke's column?

Roof plate

Central autonomic area

Central canal

Ependyma

Floor plate

Lateral funiculus

thigh?

hip?

Ventral horn
interneurons

Ventral gray
commissure

Ventral white
commissure

lower
trunk?

Ventral horn
motoneuron
columns

Ventral funiculus

Ventral median fissure

PLATE 36A

CR 108 mm
GW 14
Y68-65
Lumbar Enlargement
Cell body stain

Areas (mm²)	
Central canal	.0059
Neuroepithelium	.0118
Roof plate	.0271
Floor plate	.0056
Gray matter	2.5458
White matter	1.2621

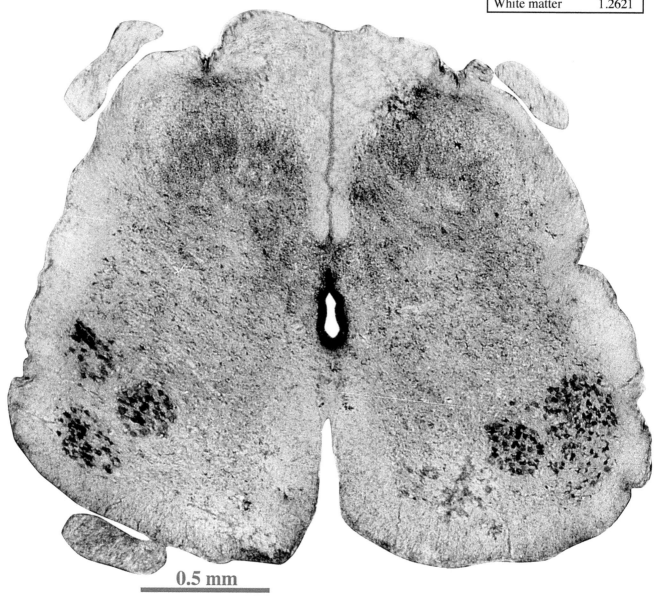

0.5 mm

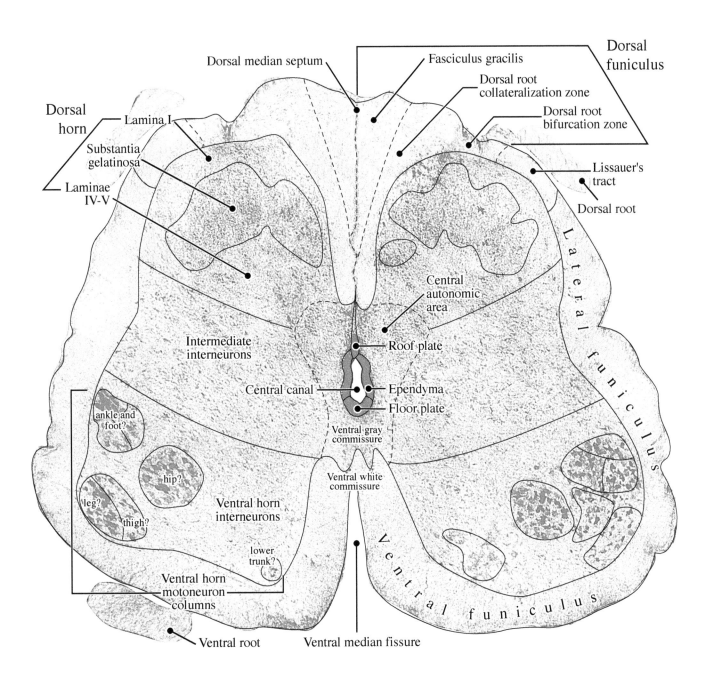

Dorsal median septum

Fasciculus gracilis

Dorsal funiculus

Dorsal root collateralization zone

Dorsal root bifurcation zone

Dorsal horn

Lamina I

Substantia gelatinosa

Laminae IV-V

Lissauer's tract

Dorsal root

Central autonomic area

Intermediate interneurons

Roof plate

Central canal

Ependyma

Floor plate

Ventral gray commissure

ankle and foot?

hip?

leg?

thigh?

Ventral horn interneurons

lower trunk?

Ventral white commissure

Ventral horn motoneuron columns

Ventral root

Ventral median fissure

Lateral funiculus

Ventral funiculus

PLATE 37A

CR 108 mm
GW 14
Y68-65
Sacral/Coccygeal
Cell body stain

Areas (mm^2)	
Central canal	.0045
Neuroepithelium	.0134
Roof plate	.0254
Floor plate	.0057
Gray matter	1.1322
White matter	.4258

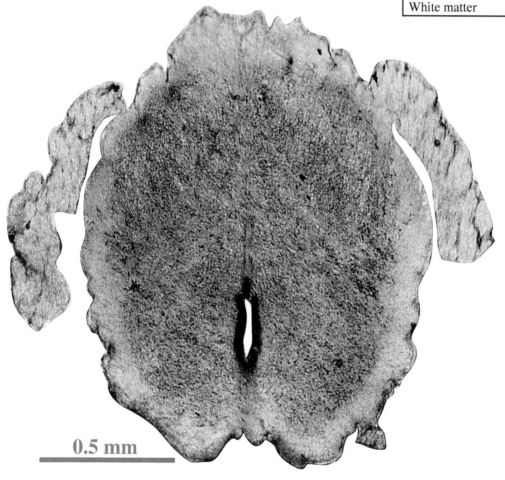

0.5 mm

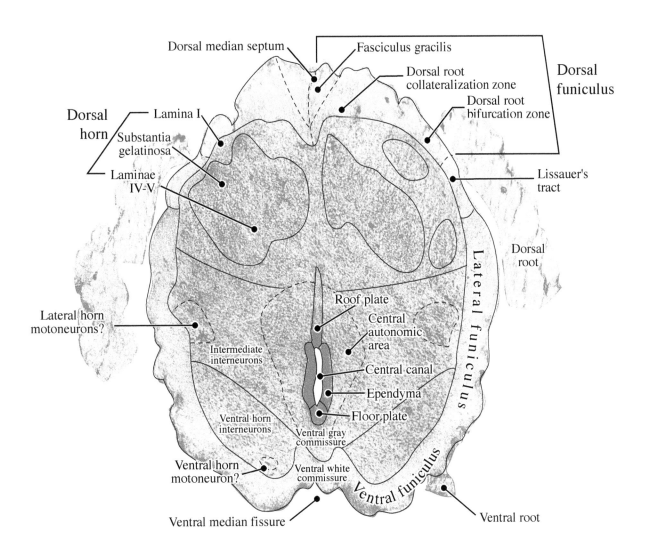

Dorsal median septum

Fasciculus gracilis

Dorsal root collateralization zone

Dorsal funiculus

Dorsal horn

Lamina I

Substantia gelatinosa

Laminae IV-V

Dorsal root bifurcation zone

Lissauer's tract

Dorsal root

Lateral funiculus

Roof plate

Central autonomic area

Central canal

Ependyma

Floor plate

Lateral horn motoneurons?

Intermediate interneurons

Ventral horn interneurons

Ventral gray commissure

Ventral horn motoneuron?

Ventral white commissure

Ventral median fissure

Ventral funiculus

Ventral root

PART VI:
3-D Reconstructions of the Cervical Level of Eight First-Trimester Specimens

Figures 4-11 feature computer reconstructed three-dimensional models of the cervical part of the developing spinal cord in 8 first-trimester specimens ranging from crown rump (CR) lengths of 3.3 mm to 56 mm (*see* **Table 1**). It is during this period that the spinal cord shows the most dramatic morphogenetic changes, starting out as a structure containing mostly neuroepithelium, and ending with a structure where most components of the neuroepithelium have disappeared, neurons have been generated, migration to the gray matter is largely completed, and most major components of the white matter appear and fill with growing axons.

Table 2 lists the colors used to distinguish thirteen different structures in the developing spinal cord in all the figures. The reconstructed models in the figures are all shown in the same orientation but differ in the number of structures that are visible to demonstrate various features of spinal cord morphogenetic development. These figures demonstrate three important relationships during spinal cord development.

First, before the appearance of a population of neurons in the gray matter, the neuroepithelial stem cells that will produce that population proliferate and expand. Those stem cells recede and ultimately disappear when all the neurons in that population are generated.

Second, there is a ventral (first) to dorsal (last) gradient of neuroepithelial growth and decline. Ventral horn neurons, especially the large alpha motoneurons, are generated earlier than neurons in the intermediate gray and the latest neurons to be generated are in the dorsal horn. These morphogenetic observations confirm the neurogenetic gradient that has been well documented in the rat spinal cord with ^{3}H–thymidine autoradiography (Altman and Bayer, 1984, 2001).

Third, the sequential neuroepithelial gradient is followed by a sequential gradient of morphogenesis. The ventral horn appears first, followed by the intermediate gray and finally by the dorsal horn.

Table 1: Specimens Used for 3-D Reconstructions of the Cervical Spinal Cord

NAME	CR	GW
C6144	3.3-mm	3.5
C836	4.0-mm	4.0
M2065	8.0-mm	6.0
C6517	10.5-mm	6.5
C8965	19.1-mm	8.0
C8553	23.0-mm	8.4
M2050	36.0-mm	10.0
Y380-62	56.0-mm	11.9

Table 2: Color Key for Structures in the Spinal Cord

STRUCTURE	COLOR
BOTH SIDES OR IN MIDLINE:	
Entire outside edge	transparent clear
Roof plate	brown
Floor plate	brown
Dorsal funiculus	pale blue
Ventral funiculus	pale violet
Lateral funiculus	pale purple
LEFT SIDE:	
Undivided neuroepithelium	dark cyan
Gray matter	bright cyan
RIGHT SIDE:	
Ventral neuroepithelium	dark red
Ventral horn	bright red
Intermediate neuroepithelium	dark yellow-green
Intermediate gray	bright yellow-green
Dorsal neuroepithelium	dark yellow
Dorsal horn	bright yellow

84

FIGURE 4   Overview of Spinal Cord Development – GW 3.5 to GW 11.9

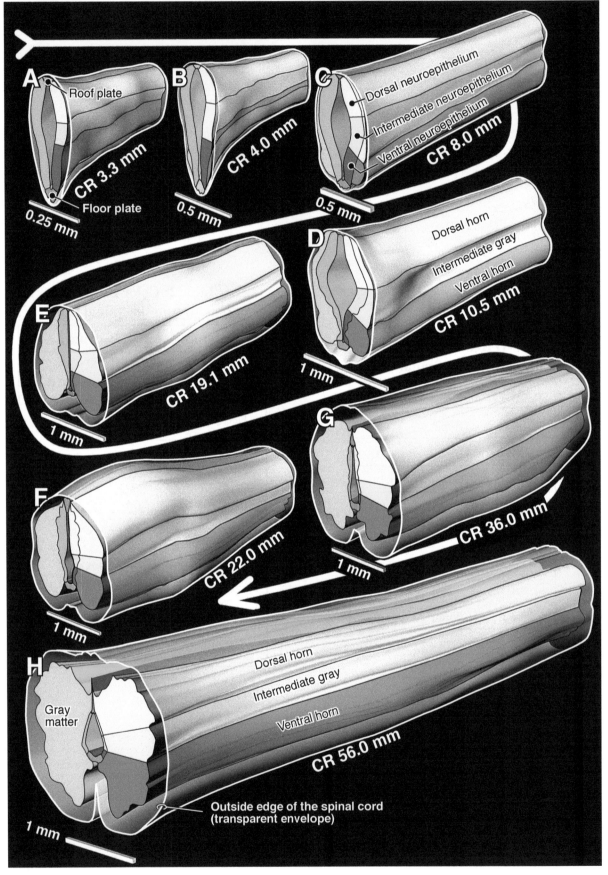

FIGURE 4

The neuroepithelium, gray matter, and the outer edge of the spinal cord are visible in all 8 specimens. The specimens are not shown to scale because structures in the smaller ones (**A-D**) cannot be seen clearly after reduction to the scale of the largest one (**H**).

At CR 3.3 mm (**A**), there is no gray matter. The ventral neuroepithelium is larger than the intermediate or dorsal neuroepithelia.

At CR 4.0 mm (**B**), a thin sliver of gray matter, the ventral horn, appears adjacent to the ventral neuroepithelium.

At CR 8.0 mm (**C**), there is gray matter adjacent to all parts of the neuroepithelium, but the ventral horn is largest. The intermediate and dorsal components of the neuroepithelium are growing larger.

At CR 10.5 mm (**D**), the ventral neuroepithelium starts to recede, but the intermediate and dorsal components are still large. The ventral horn is still the largest gray matter component, but neurons are also accumulating in the intermediate gray and dorsal horn.

That same process continues at CR 19.1 mm (**E**). The intermediate neuroepithelium is receding at CR 22.0 mm (**F**) but the dorsal neuroepithelium is still prominent. The dorsal horn is now larger than the intermediate gray.

By CR 36 mm (**G**), the dorsal neuroepithelium considerably recedes and the dorsal horn is as large as the ventral horn.

By CR 56 mm (**H**), the neuroepithelium has been replaced by an ependyma surrounding the shrinking central canal, and the dorsal and ventral horns are prominent components of the gray matter.

86

FIGURE 5

The Neuroepithelium – GW 3.2 to GW 4.0

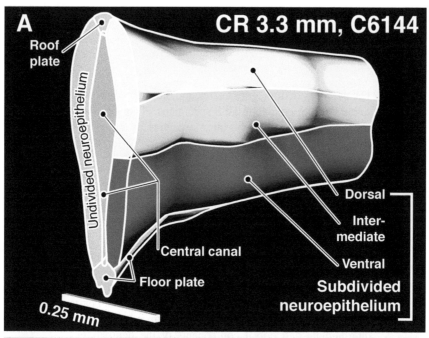

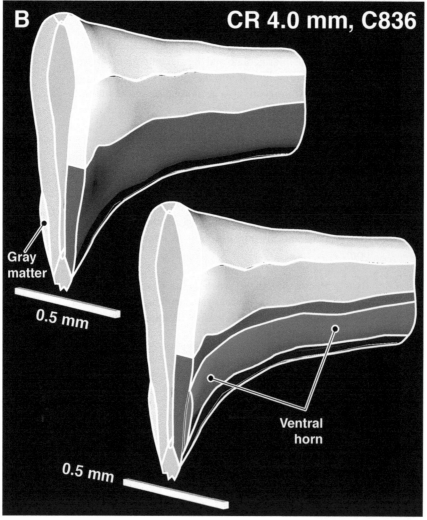

FIGURE 5

An enlarged view of the specimens at CR 3.3 mm (**A**) and CR 4.0 mm (**B**) with only the neuroepithelium, roof plate, floor plate, and gray matter visible (note scale differences).

The specimen in **A** has no gray matter. This indicates that postmitotic neurons have not yet been generated. The neuroepithelium is full of mitotically active stem cells that will eventually produce neurons.

The **top** reconstruction in **B** does not show any gray matter on the right side. The **bottom** view in **B** shows a sliver of postmitotic neurons, presumably motoneurons, adjacent to the *central part* of the ventral neuroepithelium.

Notice that the surface of the neuroepithelium is smooth in both specimens. In addition, the ventral neuroepithelium is larger than intermediate or dorsal neuroepithelia in both specimens.

FIGURE 6

Neuroepithelium and Gray Matter – GW 6 to GW 6.4

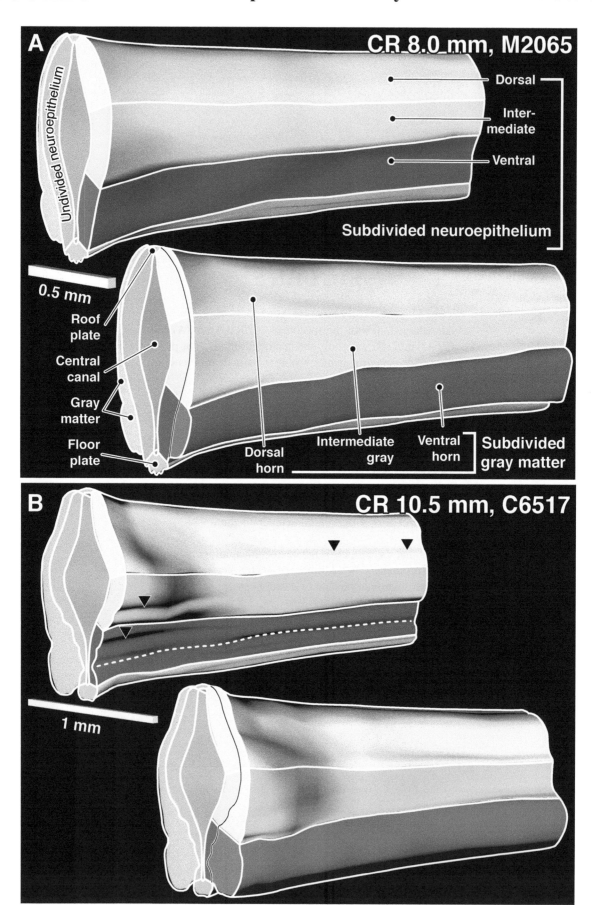

A — CR 8.0 mm, M2065

Undivided neuroepithelium

Dorsal
Inter-mediate
Ventral

Subdivided neuroepithelium

0.5 mm

Roof plate
Central canal
Gray matter
Floor plate

Dorsal horn
Intermediate gray
Ventral horn

Subdivided gray matter

B — CR 10.5 mm, C6517

1 mm

FIGURE 6

Enlarged views of the CR 8.0 mm (**A**) and CR 10.5 mm (**B**) specimens (note different scales). The **top** views do not show the gray matter on the right side so that surface features of the neuroepithelium can be observed.

In **A**, the three components of the neuroepithelium are roughly the same size, and their outer surfaces are smooth. The outer surface represents the *basal* aspect of the neuroepithelium; the *apical* aspect (inner surface) forms the edge of the central canal. The ventral horn bulges outward. There is an intermediate gray that is thicker at its junction with the ventral horn and thinner at its junction with the few postmitotic neurons accumulating in the dorsal horn. The early dorsal horn neurons may be the large Waldeyer cells (Altman and Bayer, 2001).

In **B**, the ventral neuroepithelium is beginning to recede in comparison with the still growing intermediate and dorsal neuroepithelia. The **dashed line** indicates a depression in the ventral neuroepithelium that may indicate the region where the greatest concentration of stem cells of large motoneurons are located at earlier stages. Possibly, this depression in the neuroepithelium signals the end of the neurogenetic period for motoneuron generation.

The **downward arrowheads** point to undulations in the outer surface of the intermediate and dorsal parts of the neuroepithelium that may be "sojourn zones" of postmitotic neurons. Short survival ^{3}H–thymidine autoradiographic studies in the rat spinal cord show no label uptake in these zones, even though cell density indicates that they are still within the neuroepithelium (Altman and Bayer, 2001). It is postulated that the sojourn zones are accumulations of premigratory neurons destined to settle within the intermediate gray, dorsal horn, and possibly the ventral horn. All components of the gray matter are growing larger when **B** is compared to **A**.

The ventral horn is still the most prominent gray matter component and features a smooth outer surface. There is a shallow concave area in the ventral part of the outer surface of the dorsal horn. That shallow depression is the region where axons from large cells in the dorsal root ganglion accumulate in the oval bundle of His, or what we call the dorsal root bifurcation zone (see **Figure 11**).

90

FIGURE 7

Neuroepithelium and Gray Matter – GW 8 to GW 8.4

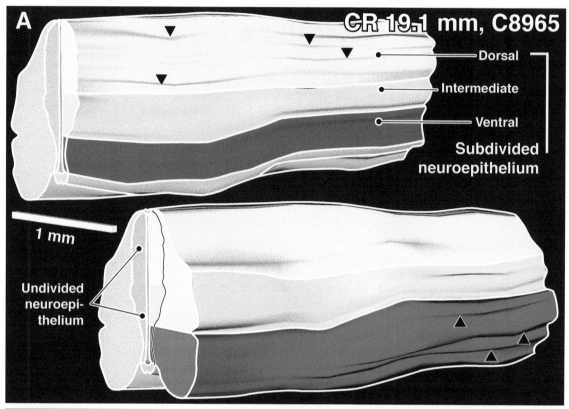

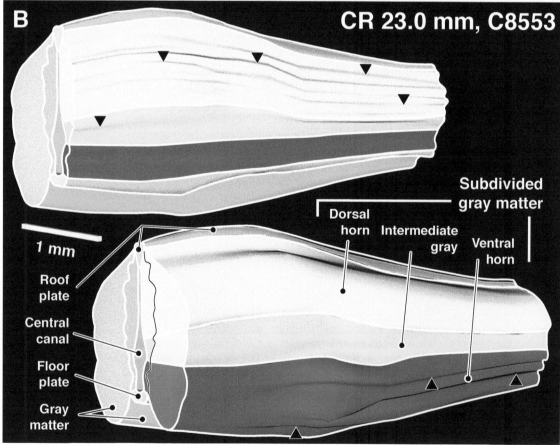

FIGURE 7

Enlarged views of the CR 19.1 mm (**A**) and CR 23 mm (**B**) specimens (note scale differences). The **top** views show only the surface features of the neuroepithelium on the right side; **bottom** views show the adjacent gray matter on the right side.

In **A**, the ventral neuroepithelium has receded but the intermediate and dorsal neuroepithelia are still active (now with many undulations, **downward arrowheads**). The outer surface of the posterior ventral horn has bulges and depressions (**upward arrowheads**) representing the segregation of the ventral motoneurons into columns in the cervical enlargement.

In **B**, the same features are present, but now the intermediate neuroepithelium is starting to recede, and the dorsal horn is growing rapidly.

FIGURE 8

Neuroepithelium and Gray Matter – GW 10 to GW 11.9

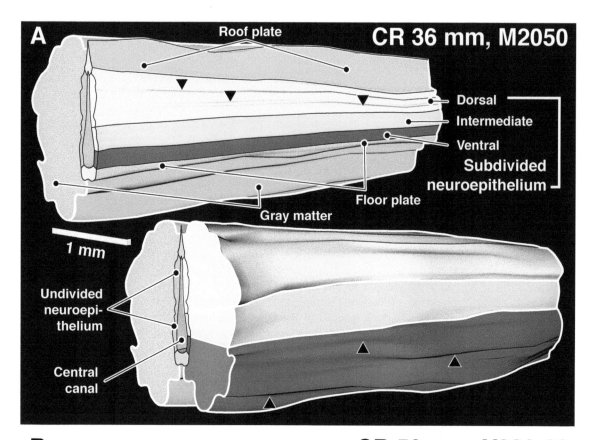

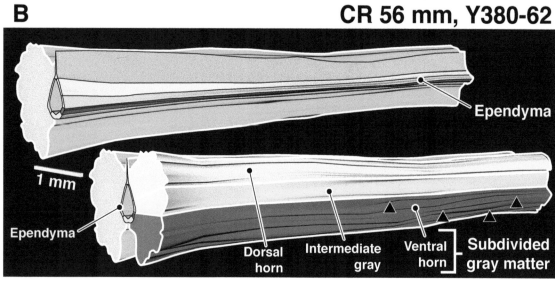

FIGURE 8

Enlarged views of the CR 36 mm (**A**) and CR 56 mm (**B**) specimens (note scale differences). The **top** views show only the surface features of the neuroepithelium on the right side; **bottom** views show the adjacent gray matter on the right side.

In **A**, all parts of the neuroepithelium are relatively smaller than in previous specimens but the dorsal neuroepithelium (still with many undulations, **downward arrowheads**) is active. The bulges in the outer surface of the ventral horn (**upward arrowheads**) extend throughout its entire length representing the advancing segregation of the ventral motoneurons into columns at all cervical levels.

In **B**, the neuroepithelium has been transformed into an ependyma that surrounds a considerably smaller central canal. The outer surface of the ependyma is smooth. The dorsal horn has grown relatively larger and reaches the same size as the ventral horn.

FIGURE 9 Neuroepithelium, Roof and Floor Plates – GW 3.5 to GW 6.5

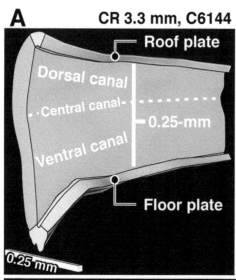

A — CR 3.3 mm, C6144

Roof plate
Dorsal canal
·Central canal-
0.25-mm
Ventral canal
Floor plate
0.25 mm

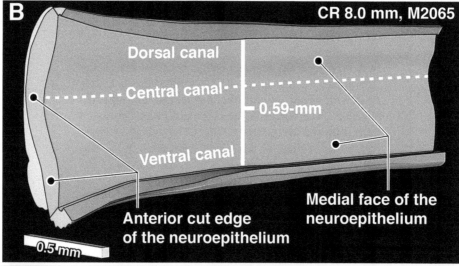

B — CR 8.0 mm, M2065

Dorsal canal
Central canal- -
0.59-mm
Ventral canal
Anterior cut edge of the neuroepithelium
Medial face of the neuroepithelium
0.5 mm

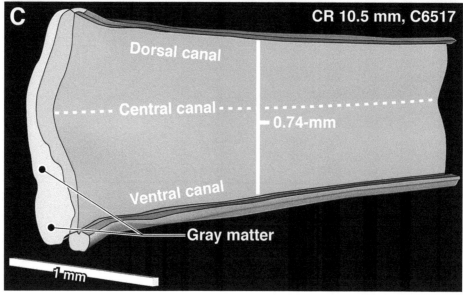

C — CR 10.5 mm, C6517

Dorsal canal
Central canal - - - -
0.74-mm
Ventral canal
Gray matter
1 mm

FIGURE 9

The undivided neuroepithelium, roof plate, floor plate, and gray matter on the left side of the spinal cord in the following specimens:

CR 3.3 mm (**A**), CR 8 mm (**B**), and CR 10.5 mm (**C**). Note the scale differences between specimens. The **dashed lines** run through the sulcus limitans.

The sulcus limitans is closer to the roof plate in **A** and shifts downward in **B** and **C** because of expansion of the dorsal neuroepithelium.

In these specimens, the roof plate is the uppermost structural feature of the spinal cord, and is always above the level of the dorsal horn. In addition, the roof plate has a smooth domed surface with no indication of a cleft in the midline.

The **vertical bars** show the actual distance (mm) between the roof and floor plates in the approximate center section of each model. The distances continually increase from **A** to **C** indicating that the neuroepithelium has a net growth during this period, even though many young neurons have already been generated and their stem cells are no longer in the neuroepithelium.

FIGURE 10   Neuroepithelium, Roof and Floor Plates – GW 7.0 to GW 14

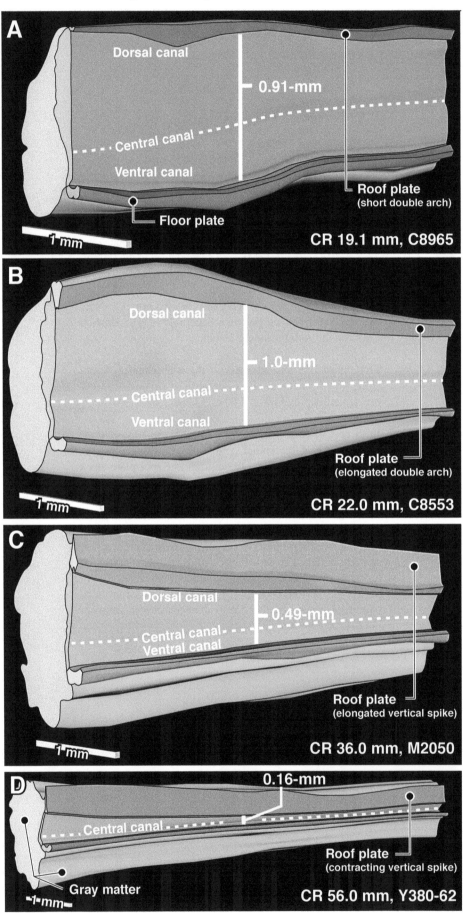

A
Dorsal canal
0.91-mm
Central canal
Ventral canal
Roof plate
(short double arch)
Floor plate
1 mm
CR 19.1 mm, C8965

B
Dorsal canal
1.0-mm
Central canal
Ventral canal
Roof plate
(elongated double arch)
1 mm
CR 22.0 mm, C8553

C
Dorsal canal
0.49-mm
Central canal
Ventral canal
Roof plate
(elongated vertical spike)
1 mm
CR 36.0 mm, M2050

D
0.16-mm
Central canal
Roof plate
(contracting vertical spike)
Gray matter
1 mm
CR 56.0 mm, Y380-62

FIGURE 10

A continuation of the series in the following specimens: CR 19.1 mm (**A**), CR 22 mm (**B**), CR 36 mm (**C**), and CR 56 mm (**D**). Note the scale differences between specimens.

The sulcus limitans (**dashed line**) moves closer and closer to the floor plate from **A** to **D**, reflecting the regression of the ventral and intermediate neuroepithelia. On the other hand, the dorsal neuroepithelium is expanding between **A** and **B**, indicating that it is still producing neurons; dorsal neuroepithelium regresses between **B** and **C**. In **D**, only the central canal remains, marking the original position of the sulcus limitans.

Note that the distance between the roof and floor plates (**vertical bars** with measurements in millimeters) reaches its maximum in **B** as the dorsal neuroepithelium stretches in the dorsoventral plane. The dorsal neuroepithelium dramatically shortens after CR 36 mm (**C**) and disappears in **D**. This dramatic shrinking marks the end of neurogenesis in the cervical spinal cord. At CR 56 mm, the distance between the roof and floor plates shrinks to a smaller value than the youngest specimen (**A, Figure 9**).

In **A** and **B**, the roof plate is positioned at the most dorsal part of the gray matter. It sinks downward in **C** as the dorsal gray matter grows because neurons settle and start to differentiate. In **D**, the massive growth of the dorsal horn causes the roof plate to sink downwards even more.

The roof plate changes its profile during its downward movement. In all panels of **Figure 7**, the roof plate surface is a single arch spanning the midline. In **A** of this series, the surface is a shallow double arch with a cleft in the midline. The roof plate still has a double arch in the dorsal midline in **B**, but there is a noticeable lengthening in the dorsoventral plane. In **C** and **D**, the double arch is replaced by a sharp vertical spike that elongates in the midline.

98

FIGURE 11

Growth of the White Matter – GW 6 to GW 11.9

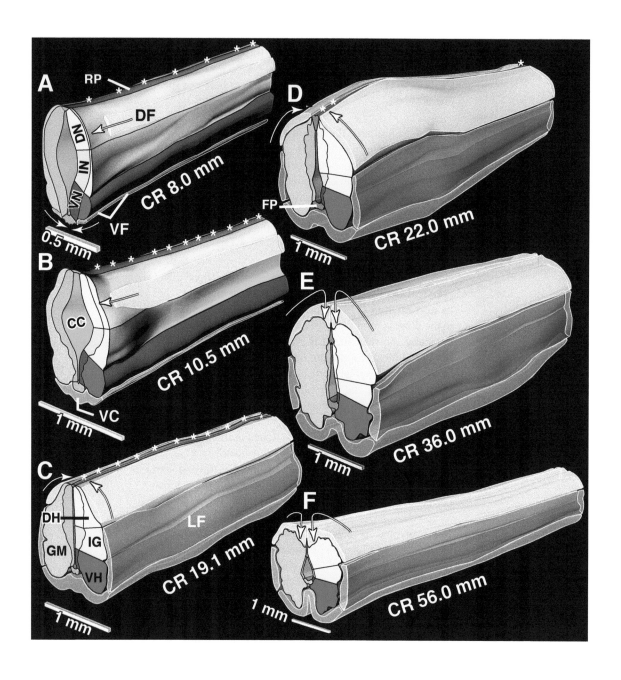

FIGURE 11

The emergence and growth of the white matter in the following specimens: CR 8 mm (**A**), CR 10.5 mm (**B**), CR 19.1 mm (**C**), CR 22 mm (**D**), CR 36 mm (**E**), and CR 56 mm (**F**). Note scale differences between specimens. All parts of the gray matter and neuroepithelium are visible in the models. **Asterisks (A to D)** indicate where the top of the roof plate (**RP**) is visible in the dorsal midline. The dorsal funiculus (**DF**) and ventral funiculus (**VF**) appear simultaneously in **A**; the lateral funiculus (**LF**) is definite in **C**.

The first axons in the ventral funiculus are most likely those of the early generated commissural neurons that accumulate in the space between the lateral edges of the floor plate (**FP**), the most ventral part of the ventral neuroepithelium (**VN**), and the ventromedial edge of the ventral horn (**VH**). In some sections of the model in **A**, the axons are even closer to the midline, so the ventral funiculus first has a medial direction of growth (**bottom arrows in A** pointing toward the midline). In **B**, these fibers cross the midline in the ventral commissure (**VC**), and the ventral funiculus increases in size and fills out laterally as well as medially. The ventral funiculus continues to enlarge throughout the remainder of spinal cord development and contains several descending tracts (medial longitudinal fasciculus, tectospinal tract, and vestibulospinal tract) as well as some of the ascending spinothalamic tracts and the collaterals of axons from neurons throughout the ventral and intermediate gray in the intraspinal tract (also known as the propriospinal tract).

The dorsal funiculus contains axons from large neurons in the dorsal root ganglia that enter the spinal cord and bifurcate (oval bundle of His) into ascending and descending branches. That bundle is more obvious at lower than upper cervical levels (**A**, **B**) because the highest dorsal root ganglion is below the highest cervical level of the spinal cord, and axons have not yet reached there. The *forward arrows* in **A** and **B** indicate the advance of the dorsal funiculus. Axons continue to accumulate in the dorsal funiculus, and it reaches the midline in **C** and **D** (*upward arrows*). In **E** and **F**, the dorsal funiculus expands downward (*downward arows*). These axons are segregated in the midline by a band of glia that extends to the pia from the top of the roof plate (see **C** and **D** in **Figure 10**). The medial part of the dorsal funiculus contains the long ascending branches of the dorsal root axons in the fasciculus gracilis that comes from the T5 dorsal root ganglion and below. The fasciculus cuneatus contains axons from the T4 dorsal root ganglion and above (*see* Chapter 5, Figures 5-28 and 5-49 in Altman and Bayer, 2001).

PART VII:
3-D Reconstructions of the Progressive Segregation
of Ventral-Horn Motor Neurons into Columns

This section features computer-generated 3-dimensional models of the entire spinal cord in three specimens: CR 36 mm (GW 10), CR 56 mm (GW 11.9), and CR 108 mm (GW 14). Throughout the ventral horn, individual sections contain clumps of motoneurons. When the sections are aligned, these clumps form *longitudinal columns*. Each distinguishable motoneuron column has been reconstructed in the three specimens. **Figures 12–13** provide an overview of the sections included in the GW 10 model; the model is shown in **Figures 12–15**. **Figures 16–18** provide an overview of the sections included in the GW 11.9 model; the model is shown in **Figures 19–22**. **Figures 23–25** provide an overview of the sections included in the GW 14 model; the model is shown in **Figures 26–31**. The segregation of motoneuron columns is a progressive developmental process. On each side of the ventral horn, there are 7 motor columns in the GW 10 specimen, 9 in the GW 11.9 specimen, and 15 in the GW 14 specimen.

Experimental studies in animals show that motoneuron segregation is related to dendritic differentiation (*see* Chapter 1, Section 1.6 and Chapter 3, Section 3.3 in Altman and Bayer, 2001). When there are fewer columns containing larger groups of neurons, dendrites are growing in many directions. As development progresses, the dendrites are reshaped into well-defined bundles that extend longitudinally in the spinal cord (*see* Figures 1-68, 1-69, 3-23 and 3-24 in Altman and Bayer, 2001). That process produces progressively better defined motoneuron columns in cell-body-stained sections. Possibly, motoneuron dendritic bundles develop similarly in man. Compare, for example, the progressive segregation of motoneurons in the cervical enlargement in **Plates 15, 23,** and **31**.

All parts of the spinal cord contain at least one motoneuron column located in the ventromedial part of the ventral horn. In the thoracic cord, that is the only column present. Progressively more lateral columns are found at cervical and lumbar levels, especially in their respective enlargements. Traditionally, these columns are named according to their *position*. In our book on development of the human spinal cord (Altman and Bayer, 2001) that locational terminology has been put into a grid system containing up to 4 panels (medial, central, lateral, and far-lateral) and 3 tiers (ventral, dorsal, and retrodorsal). The positional labeling system is functionally neutral, because little is known about the exact groups of muscles that are innervated by motoneurons in the human spinal cord. The overview figures that introduce each model contain the positional labeling for each reconstructed motoneuron column.

However, there is a wealth of experimental anatomical evidence in animals, including primates, that the locations of motoneuron columns in the ventral horn have functional significance (*see* Chapter 1, Section 1.6 and Chapter 2, Section 2.3.2 in Altman and Bayer, 2001). The ventromedial columns at all levels (medial panel, ventral tier) contain motoneurons that innervate the axial muscles. More laterally placed columns at upper cervical and upper lumbar levels contain motoneurons that innervate proximal limb muscles (central, lateral, and far-lateral panels, ventral to dorsal tiers). Lateral columns in the cervical and lumbar enlargements contain motoneurons that innervate distal limb muscles (especially those located in the far-lateral panel and retrodorsal tier).

FIGURE 12

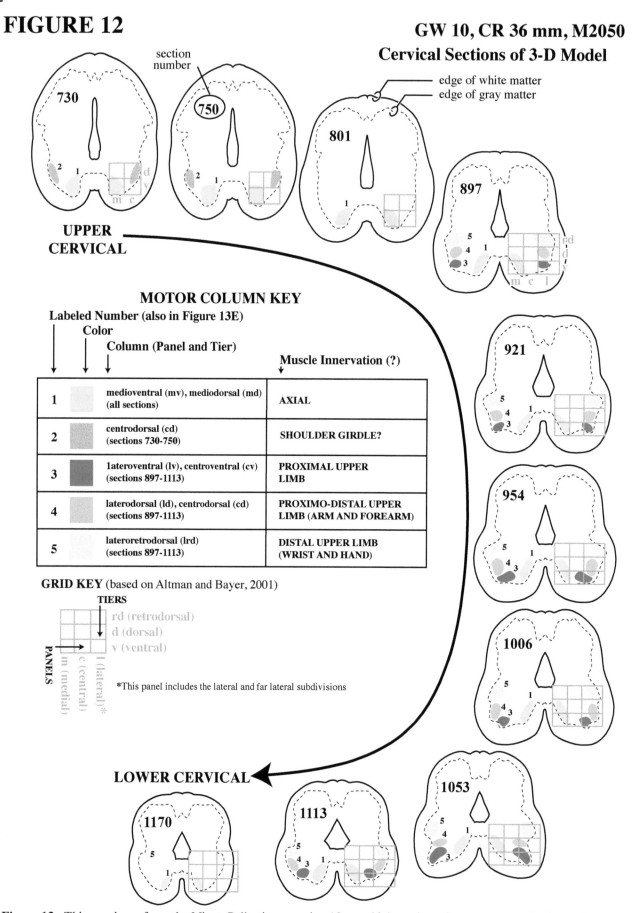

section
number

edge of white matter
edge of gray matter

730

750

801

897

UPPER CERVICAL

MOTOR COLUMN KEY

Labeled Number (also in Figure 13E)

Color

Column (Panel and Tier)

Muscle Innervation (?)

1		medioventral (mv), mediodorsal (md) (all sections)	AXIAL
2		centrodorsal (cd) (sections 730-750)	SHOULDER GIRDLE?
3		1ateroventral (lv), centroventral (cv) (sections 897-1113)	PROXIMAL UPPER LIMB
4		laterodorsal (ld), centrodorsal (cd) (sections 897-1113)	PROXIMO-DISTAL UPPER LIMB (ARM AND FOREARM)
5		lateroretrodorsal (lrd) (sections 897-1113)	DISTAL UPPER LIMB (WRIST AND HAND)

GRID KEY (based on Altman and Bayer, 2001)

TIERS

rd (retrodorsal)
d (dorsal)
v (ventral)

PANELS

m (medial)
c (central)
l (lateral)*

*This panel includes the lateral and far lateral subdivisions

921

954

1006

LOWER CERVICAL

1053

1113

1170

Figure 12. This specimen from the Minot Collection contains 10-μm-thick sections that are consecutively numbered on the slides. All outlines are profiles of the sections used in the cervical spinal cord. Three of these sections are illustrated in the Atlas. **Section 730** is in **Plate 14**; **section 921** is in **Plate 15**, and **section 1113** is in **Plate 16**. After fixation, the total length of the cervical part of the model (shown in **Figure 13**) extends 4.4 mm.

GW 10, CR 36 mm, M2050

FIGURE 13

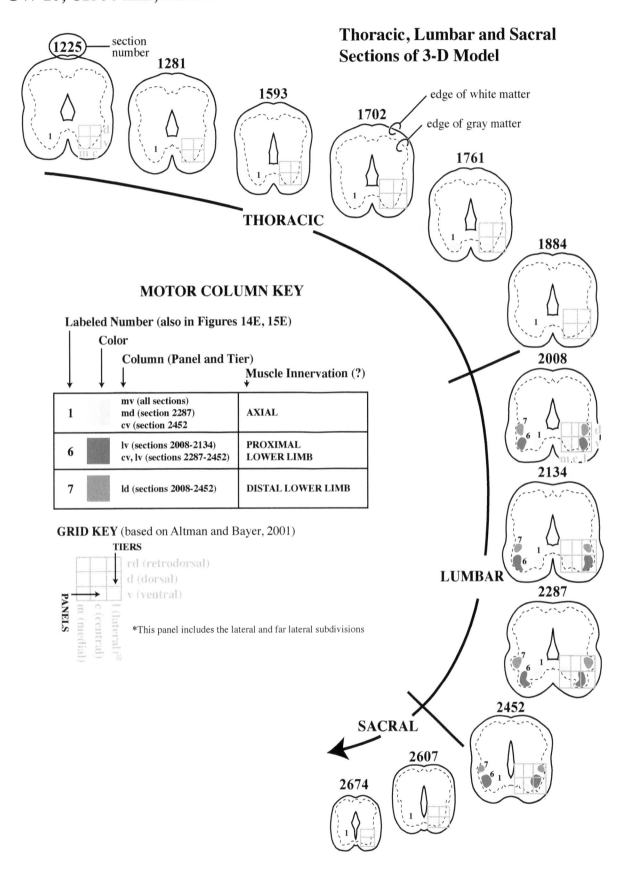

Figure 13. A continuation of the previous figure showing outlines of the sections used in the thoracic, lumbar, and sacral spinal cord. Six of these sections are illustrated in the Atlas. **Section 1593** is in **Plate 17**; **section 1884** is in **Plate 18**, **section 2134** is in **Plate 19,** and **sections 2607** and **2674** are in **Plate 20**. After fixation, the total length of the thoracic part of the model (**Figure 14**) is 6.99 mm, the combined lumbar and sacral regions is 6.66 mm (**Figure 15**).

FIGURE 14

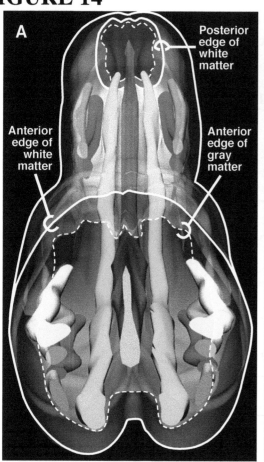

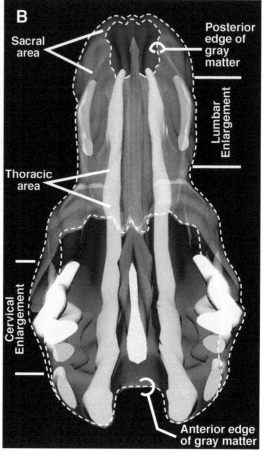

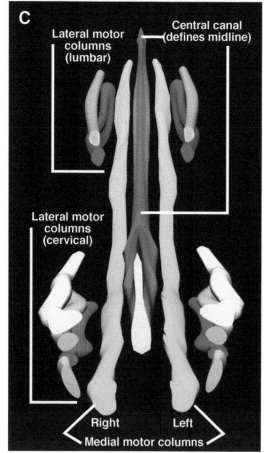

Figure 14

The entire spinal cord in specimen M2050 (GW 10, CR 36 mm) as viewed from the front and top (**A, B, C,** this page) or the upper right side (**D, E, F**, facing page). The front edge of the reconstruction is section 730, the back edge is section 2674. The total length after fixation is 19.44 mm. In this and the following reconstructions, the spinal cord has been enlarged 3 times more crosswise than lengthwise. Otherwise, the model is too long to see the various motoneuron columns clearly.

Solid outlines define the outer edges of the white matter (transparent envelope, **A** and **D**); **dashed outlines** define the outer edges of the gray matter (less transparent envelope, **B** and **E**). The **gray-white solid** in panels **A-E** is the central canal. The **colored solids** in panels **A-E** are ventral horn motor columns.

There are 7 distinguishable pairs of motor columns in this specimen. The medial motor columns (**cyan**) extend throughout the entire length of the spinal cord. There are four lateral motor columns in the cervical region, a short one (**light brown**) in the upper cervical and three (**yellow, orange, red**) in the cervical enlargement. The ventral to dorsal stacking of the lateral motor columns are more easily observed in the side views (especially panel **F**). At this stage of development, the lumbar enlargement is not prominent, and contains only two lateral motor columns (**purple and pink**).

GW 10, CR 36 mm, M2050, Side View

FIGURE 14

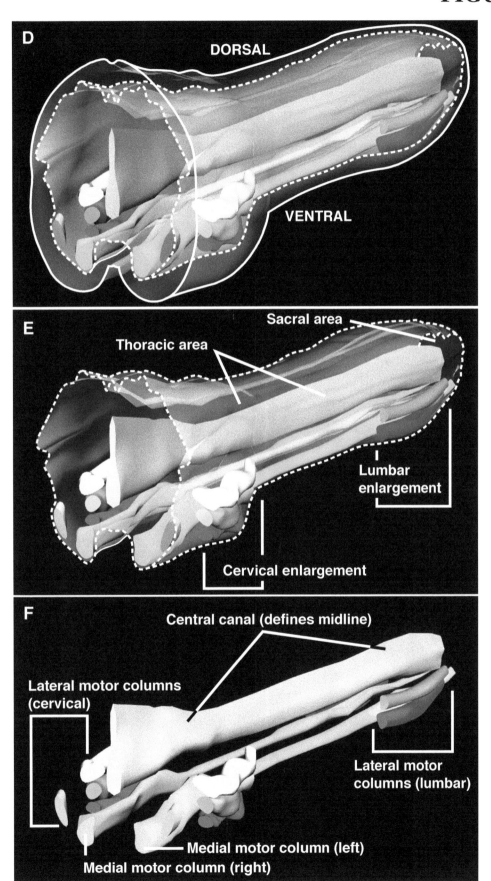

106

FIGURE 15

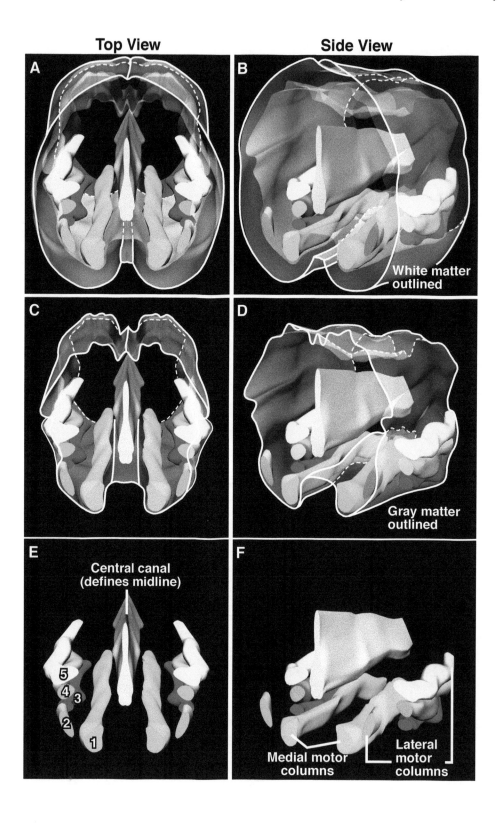

Top View

Side View

A

B White matter outlined

C

D Gray matter outlined

E Central canal (defines midline)
5 4 3 2 1

F Medial motor columns   Lateral motor columns

FIGURE 15

The cervical region in specimen M2050 (GW 10, CR 36 mm). The individual sections (730–1170) in this model are diagrammed in **Figure 12**. The reconstructed length in the specimen is 4.4 mm after fixation.

Panels **A**, **C**, and **E** show the model from the top front, pancls **B**, **D**, and **F** from the upper right side. Panels **A** and **B** show both the white matter (**solid outlined** outer transparent envelope) and gray matter (**dashed-outlineed** inner transparent envelope) around the central canal (**gray-white solid**) and motor columns (**colored solids**). Panels **C** and **D** show the only gray matter (**dashed** and **solid outlined** outer transparent envelope) around the central canal and motor columns. Panels **E** and **F** show the central canal and motor columns alone.

There are **5 motor columns**:
1 (**cyan**, axial muscles?) is located in the medial panel, ventral and dorsal tiers.
2 (**light brown**, shoulder motoneurons?) is located in the central panel, dorsal tier at high cervical levels.
3 (**red**, proximal upper limb motoneurons?) in the cervical enlargement is located in the lateral and central panels, ventral tier.
4 (**orange**, arm and forearm muscle motoneurons?) in the cervical enlargement is located in the lateral and central panels, dorsal tier.
5 (**yellow**, wrist and hand muscle motoneurons?) in the cervical enlargement is located in the lateral panel, retrodorsal tier.

FIGURE 16

Top View **Side View**

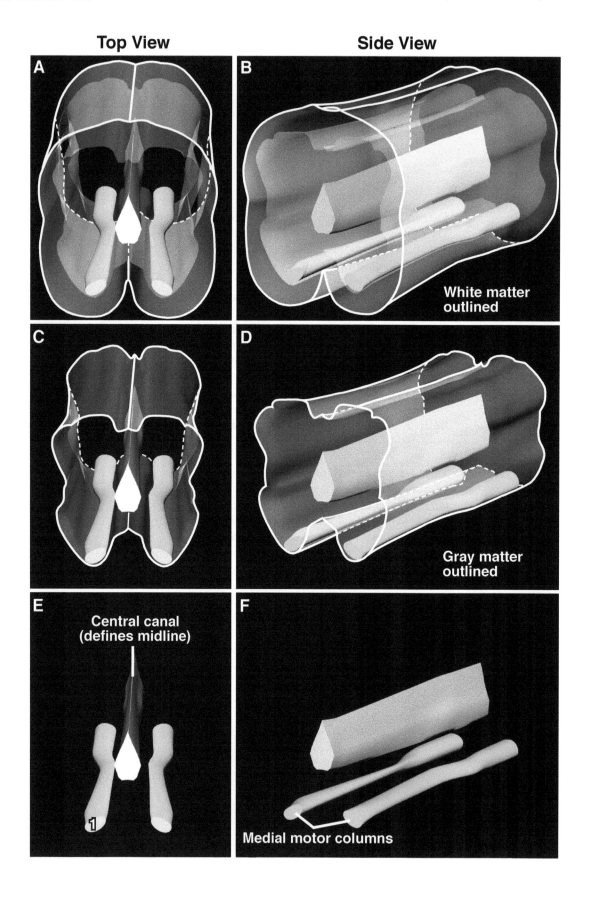

A

B

White matter
outlined

C

D

Gray matter
outlined

E

Central canal
(defines midline)

1

F

Medial motor columns

FIGURE 16

The thoracic region in specimen M2050 (GW 10, CR 36 mm). The individual sections in this model are diagrammed in **Figure 13**, and include sections 1225-1884. The reconstructed length in the specimen is 6.99 mm after fixation.

Panels **A**, **C**, and **E** show the model from the top front, panels **B**, **D**, and **F** from the upper-right side. Panels **A** and **B** show both the white matter (**outlined** outer transparent envelope) and gray matter (inner transparent envelope) around the central canal (**gray-white solid**) and the medial motor columns (**cyan**). Panels **C** and **D** show the only gray matter (**outlined** outer transparent envelope) around the central canal and the medial motor columns. Panels **E** and **F** show the central canal and the medial motor columns alone.

This region of the spinal cord is distinguished from other regions by having only one motor column (**1**) that probably innervates axial muscles associated with the thoracic vertebrae, rib cage, and dorsal abdominal wall.

FIGURE 17

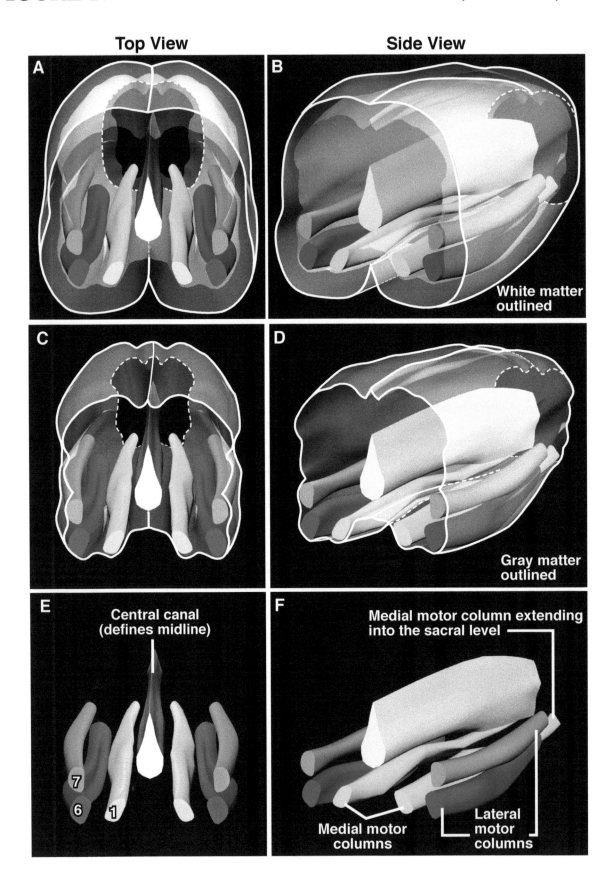

Top View

Side View

A

B — White matter outlined

C

D — Gray matter outlined

E — Central canal (defines midline)

7   6   1

F — Medial motor column extending into the sacral level

Medial motor columns   Lateral motor columns

FIGURE 17

The lumbar and sacral regions in specimen M2050 (GW 10, CR 36 mm). The individual sections in this model are diagrammed in **Figure 11**, and include sections 2008–2674. The reconstructed length in the specimen is 6.66 mm after fixation.

Panels **A**, **C**, and **E** show the model from the top front, panels **B**, **D**, and **F** from the upper-right side. Panels **A** and **B** show both the white matter (**outlined** outer transparent envelope) and gray matter (inner transparent envelope) around the central canal (**gray-white solid**) and motor columns (**colored solids**). Panels **C** and **D** show the gray matter (**outlined** outer transparent envelope) around the central canal and motor columns. Panels **E** and **F** show the central canal and motor columns alone.

There are **3 Motor Columns**:
1 (**cyan**, axial muscle motoneurons?) is located in the medial and central panels, ventral and dorsal tiers.
6 (**purple**, proximal lower limb muscle motoneurons?) in the lumbar enlargement is located in the central and lateral panels, ventral tier.
7 (**pink**, distal lower limb muscle motoneurons?) in the lumbar enlargement is located in the lateral panel, dorsal tier.

FIGURE 18

GW 11.9, CR 56 mm, Y380-62

Cervical Sections of 3-D Model

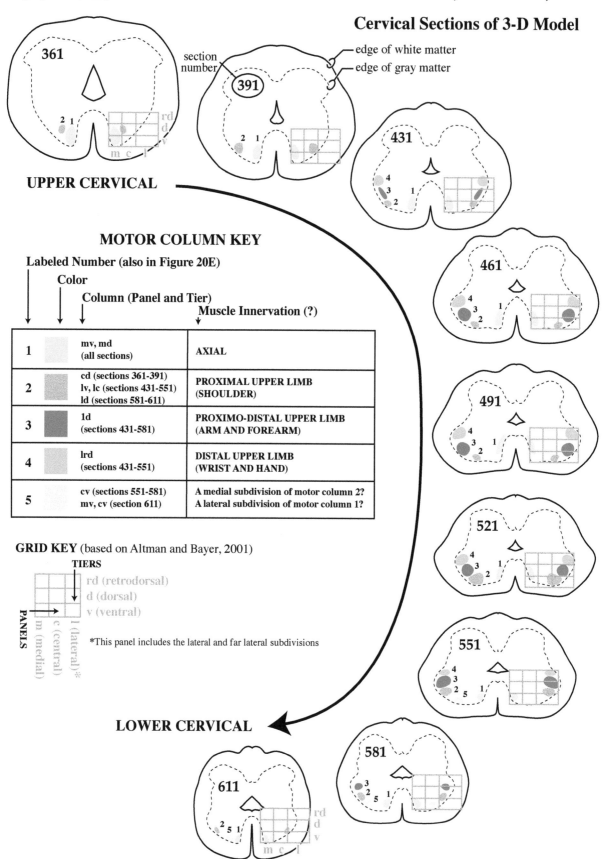

Figure 18. This specimen from the Yakovlev Collection contains serially numbered 35-μm-thick sections. All outlines are profiles of the sections used in the cervical region of the spinal cord. These profiles were constructed from the right side of the section, then copied to the left side to make the model symmetrical. After fixation, the cervical region of the model is 8.79 mm long. Two of these sections are illustrated in the Atlas. **Section 361** is in **Plate 22**; **section 521** is in **Plate 23**. The cervical model is shown by itself in **Figure 21**.

GW 11.9, CR 56 mm, Y380-62

FIGURE 19

Thoracic Sections of 3-D Model

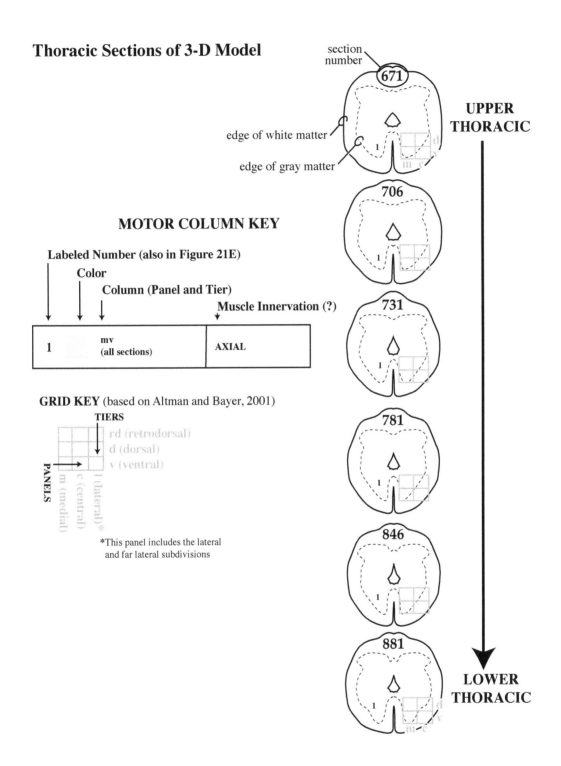

MOTOR COLUMN KEY

Labeled Number (also in Figure 21E)

Color

Column (Panel and Tier)

Muscle Innervation (?)

1	mv (all sections)	AXIAL

GRID KEY (based on Altman and Bayer, 2001)

TIERS

rd (retrodorsal)
d (dorsal)
v (ventral)

PANELS

m (medial)
c (central)
l (lateral)*

*This panel includes the lateral and far lateral subdivisions

section number

671

edge of white matter

edge of gray matter

UPPER THORACIC

706

731

781

846

881

LOWER THORACIC

Figure 19. A continuation of the previous figure showing outlines of the sections used in the thoracic region of the spinal cord. Two of these sections are illustrated in the Atlas. **Section 671** is in **Plate 24**; **section 881** is in **Plate 25**. After fixation, the total length of the thoracic part of the model is 7.35 mm. The thoracic model is shown by itself in **Figure 22**.

FIGURE 20

Lumbar and Sacral Sections of 3-D Model

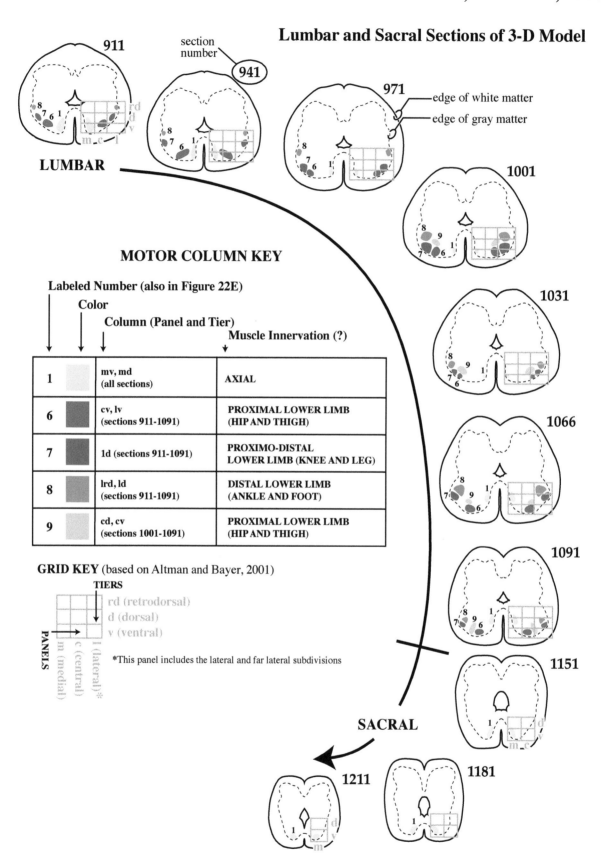

MOTOR COLUMN KEY

Labeled Number (also in Figure 22E)

Color

Column (Panel and Tier)

Muscle Innervation (?)

Labeled Number	Color	Column (Panel and Tier)	Muscle Innervation (?)
1		mv, md (all sections)	AXIAL
6		cv, lv (sections 911-1091)	PROXIMAL LOWER LIMB (HIP AND THIGH)
7		ld (sections 911-1091)	PROXIMO-DISTAL LOWER LIMB (KNEE AND LEG)
8		lrd, ld (sections 911-1091)	DISTAL LOWER LIMB (ANKLE AND FOOT)
9		cd, cv (sections 1001-1091)	PROXIMAL LOWER LIMB (HIP AND THIGH)

GRID KEY (based on Altman and Bayer, 2001)

TIERS

rd (retrodorsal)
d (dorsal)
v (ventral)

PANELS
m (medial)
c (central)
l (lateral)*

*This panel includes the lateral and far lateral subdivisions

Figure 20. A continuation of the previous figure showing outlines of the sections used in the lumbar (911-1091) and sacral (1151-1211) regions of the spinal cord. Three of these sections are illustrated in the Atlas. **Section 911** is in **Plate 26**; **section 1006** is in **Plate 27**, **section 1211** is in **Plate 28**. After fixation, the total length of the lumbosacral part of the model is 10.5 mm. The lumbosacral part of the model is in **Figure 23**.

FIGURE 21

GW 11.9, CR 56 mm, Y380-62

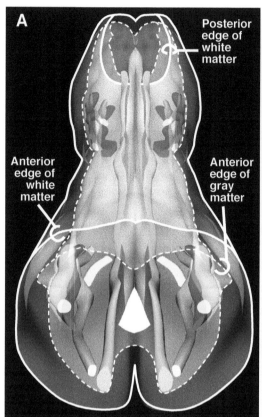

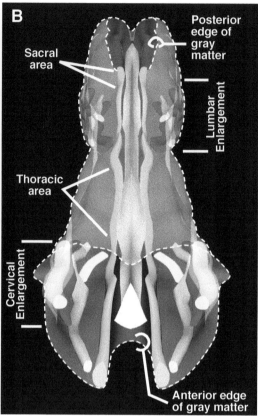

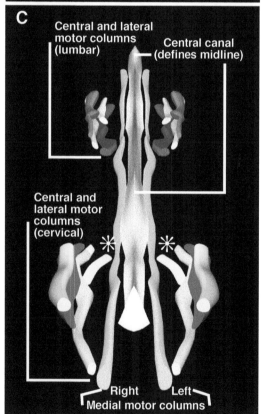

Figure 21

The entire spinal cord in specimen Y380-62 (GW 11.9, CR 56 mm) as viewed from the front and top (**A, B, C,** this page) or the upper-right side (**D, E, F,** facing page). The front edge of the reconstruction is section 361, the back edge is section 1211. The total length after fixation is 29.79 mm. In this and the following reconstructions, the spinal cord has been enlarged 3 times more crosswise than lengthwise. Otherwise, the model is too long to see the various motoneuron columns clearly.

Solid outlines define the outer edges of the white matter (transparent envelope, **A** and **D**); **dashed outlines** define the outer edges of the gray matter (less transparent envelope, **A, B** and **D, E**). The **gray-white solid** in panels **A-E** is the central canal. The **colored solids** in panels **A-E** are ventral horn motor columns. There are **9 pairs of motor columns** on each side of the ventral horn. The medial motor columns (**cyan**) extend throughout the entire length, including sacral levels. There are four lateral motor columns in the cervical region, a long one (**light brown**) extending throughout the entire region. Three (**red, orange, yellow**) are limited to the cervical enlargement. The lumbar enlargement is more prominent than at GW 10, and contains four central and lateral motor columns (**purple, pink, violet, pale violet**). The asterisks in **C** indicate the motor columns in the cervical enlargement that are not visible in the side view. The ventral to dorsal stacking of the lateral motor columns are more easily observed in the side views (especially panel **F**).

GW 11.9, CR 56 mm, Y380-62 **FIGURE 21**

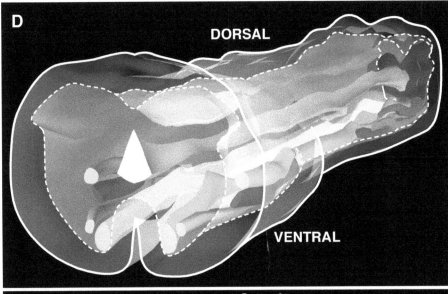

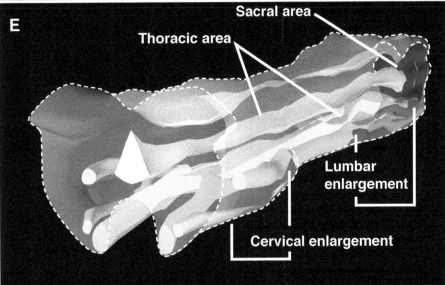

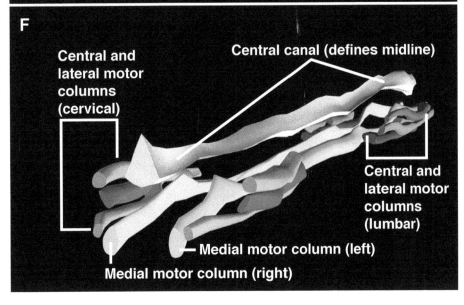

FIGURE 22

Top View

Side View

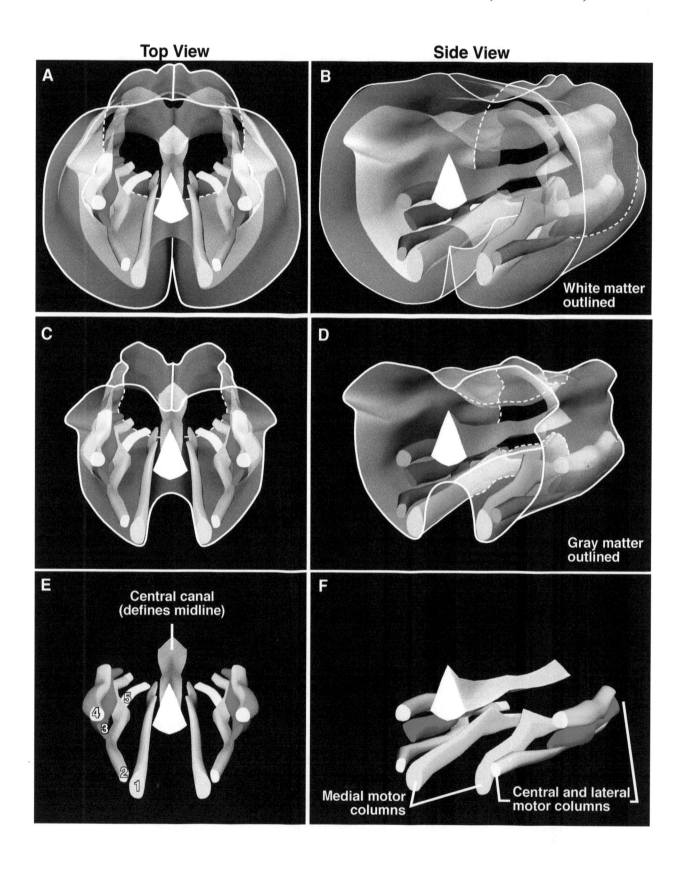

A

B

White matter outlined

C

D

Gray matter outlined

E

Central canal (defines midline)

5
4
3
2
1

F

Medial motor columns

Central and lateral motor columns

FIGURE 22

The cervical region in specimen Y380-62 (GW 11.9, CR 56 mm). The individual sections in this model are diagrammed in **Figure 18**, and include sections 361-611. The reconstructed length is 8.79 mm after fixation.

Panels **A**, **C**, and **E** show the model from the top front, panels **B**, **D**, and **F** from the upper-right side. Panels **A** and **B** show both the white matter (**outlined** transparent envelope) and gray matter (inner transparent envelope) around the central canal (**gray-white solid**) and motor columns (**colored solids**). Panels **C** and **D** show the gray matter (**outlined** transparent envelope) around the central canal and motor columns. Panels **E** and **F** show the central canal and motor columns alone.

There are **5 Motor Columns**:
1 (**cyan**, axial muscle motoneurons?) is located in the medial panel, ventral and dorsal tiers.
2 (**light brown**, proximal upper limb muscle motoneurons?) is located in the central and lateral panels, ventral and dorsal tiers.
3 (**red**, arm and forearm muscle motoneurons?) is located in the cervical enlargement in the lateral panel, dorsal tier.
4 (**orange**, wrist and hand muscle motoneurons?) is located in the cervical enlargement in the lateral panel, retrodorsal tier.
5 (**yellow**, forearm muscle motoneurons?) in the posterior cervical enlargement (sections 551-611) is located in the medial and central panels, ventral tier.

FIGURE 23

Top View

Side View

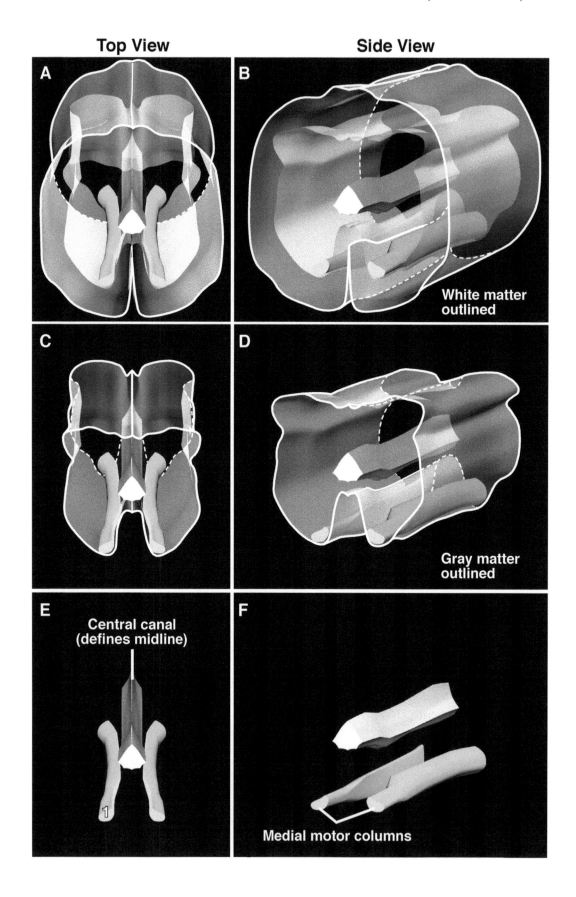

A

B — White matter outlined

C

D — Gray matter outlined

E — Central canal (defines midline) — 1

F — Medial motor columns

FIGURE 23

The thoracic region in specimen Y380-62 (GW 11.9, CR 56 mm). The individual sections in this model are diagrammed in **Figure 19**, and include sections 671-881. The reconstructed length is 7.35 mm after fixation.

Panels **A**, **C**, and **E** show the model from the top front, panels **B**, **D**, and **F** from the upper-right side. Panels **A** and **B** show both the white matter (**outlined** transparent envelope) and gray matter (inner transparent envelope) around the central canal (**gray-white solid**) and the medial motor columns (**cyan**). Panels **C** and **D** show the gray matter (**outlined** transparent envelope) around the central canal and the medial motor columns. Panels **E** and **F** show the central canal and the medial motor columns alone.

This region of the spinal cord is distinguished from other regions by having only **one motor column** (**1**) that probably innervates axial muscles associated with the thoracic vertebrae, rib cage, and dorsal abdominal wall.

FIGURE 24

Top View　　　　　　　　**Side View**

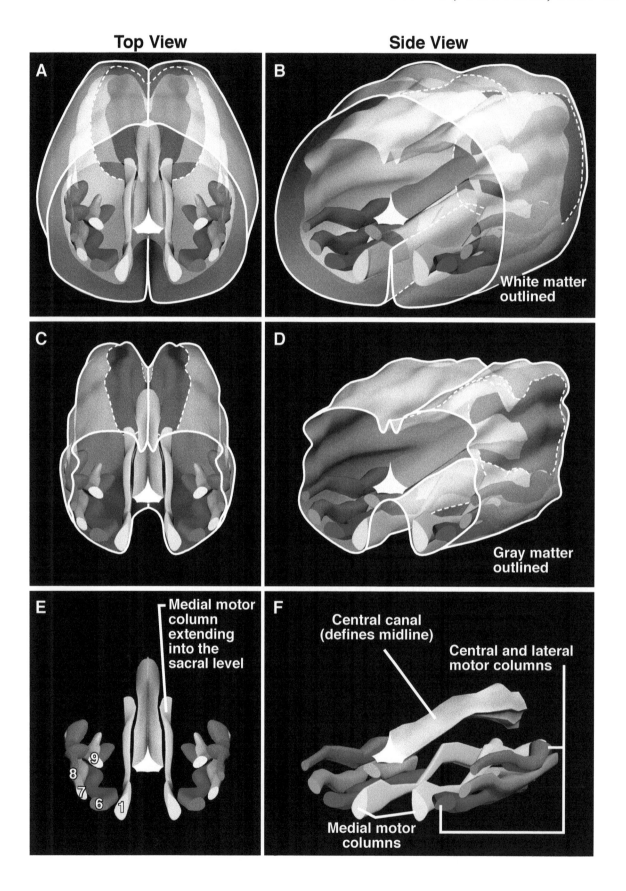

A

B White matter outlined

C

D Gray matter outlined

E Medial motor column extending into the sacral level

9
8
7
6 1

F Central canal (defines midline)　Central and lateral motor columns

Medial motor columns

FIGURE 24

The lumbar and sacral regions in specimen Y380-62 (GW 11.9, CR 56 mm). The individual sections (911-1211) in this model are diagrammed in **Figure 20**. The reconstructed length of the specimen is 10.5 mm after fixation.

Panels **A**, **C**, and **E** show the model from the top front, panels **B**, **D**, and **F** from the upper-right side. Panels **A** and **B** show both the white matter (**outlined** transparent envelope) and gray matter (inner transparent envelope) around the central canal (**gray-white solid**) and motor columns (**colored solids**). Panels **C** and **D** show the gray matter (**outlined** transparent envelope) around the central canal and motor columns. Panels **E** and **F** show the central canal and motor columns alone.

There are **5 Motor Columns**:
1 (**cyan**, axial muscle motoneurons?) is located in the medial and central panels, ventral and dorsal tiers.
6-9 are limited to lumbar levels (sections 911-1091).
6 (**dark purple**, motoneurons controlling proximal lower limb?) is located in the central and lateral panels, ventral tier.
7 (**pink**, motoneurons controlling proximo-distal lower limb?) is located in the lateral panel, dorsal tier.
8 (**violet**, motoneurons controlling distal lower limb?) is located in the lateral panel, dorsal and retrodorsal tiers.
9 (**pale violet**, motoneurons controlling proximal lower limb?) is located in the central panel, dorsal and ventral tiers.

FIGURE 25

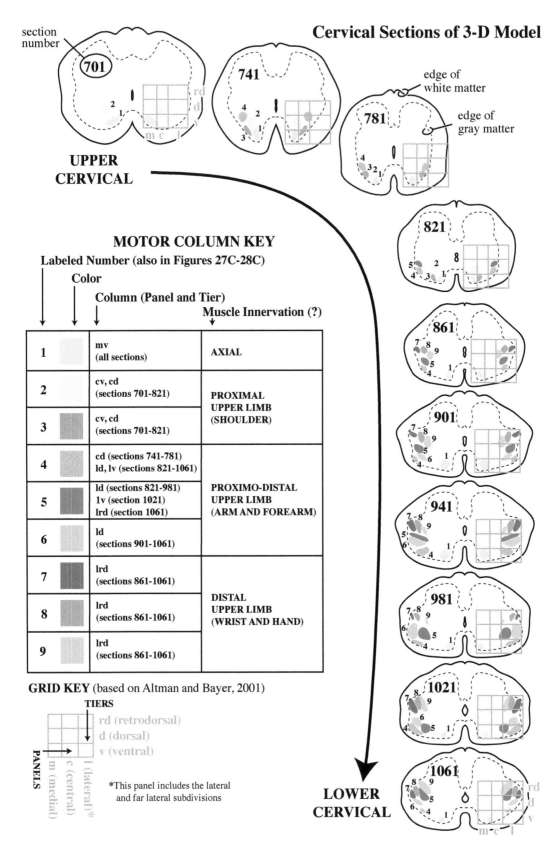

Cervical Sections of 3-D Model

section number

701

UPPER CERVICAL

741

781

edge of white matter

edge of gray matter

821

861

901

941

981

1021

1061

LOWER CERVICAL

MOTOR COLUMN KEY

Labeled Number (also in Figures 27C-28C)

Color

Column (Panel and Tier)

Muscle Innervation (?)

Labeled Number	Color	Column (Panel and Tier)	Muscle Innervation (?)
1		mv (all sections)	AXIAL
2		cv, cd (sections 701-821)	PROXIMAL UPPER LIMB (SHOULDER)
3		cv, cd (sections 701-821)	
4		cd (sections 741-781) ld, lv (sections 821-1061)	PROXIMO-DISTAL UPPER LIMB (ARM AND FOREARM)
5		ld (sections 821-981) lv (section 1021) lrd (section 1061)	
6		ld (sections 901-1061)	
7		lrd (sections 861-1061)	DISTAL UPPER LIMB (WRIST AND HAND)
8		lrd (sections 861-1061)	
9		lrd (sections 861-1061)	

GRID KEY (based on Altman and Bayer, 2001)

TIERS

rd (retrodorsal)
d (dorsal)
v (ventral)

PANELS

m (medial)
c (central)
l (lateral)*

*This panel includes the lateral and far lateral subdivisions

Figure 25. This second trimester specimen from the Yakovlev Collection contains serially numbered 35-μm-thick sections. All outlines are profiles of the sections used in the cervical region of the spinal cord. These profiles were constructed from the right side of the section, then copied to the left side to make the model symmetrical. **Section 741** is in **Plate 30**; **section 981** is in **Plate 31**. The cervical model is shown by itself in **Figures 29 and 30**. After fixation, the cervical region of the model is 12.6 mm long.

GW14, CR 108 mm, Y68-65

FIGURE 26

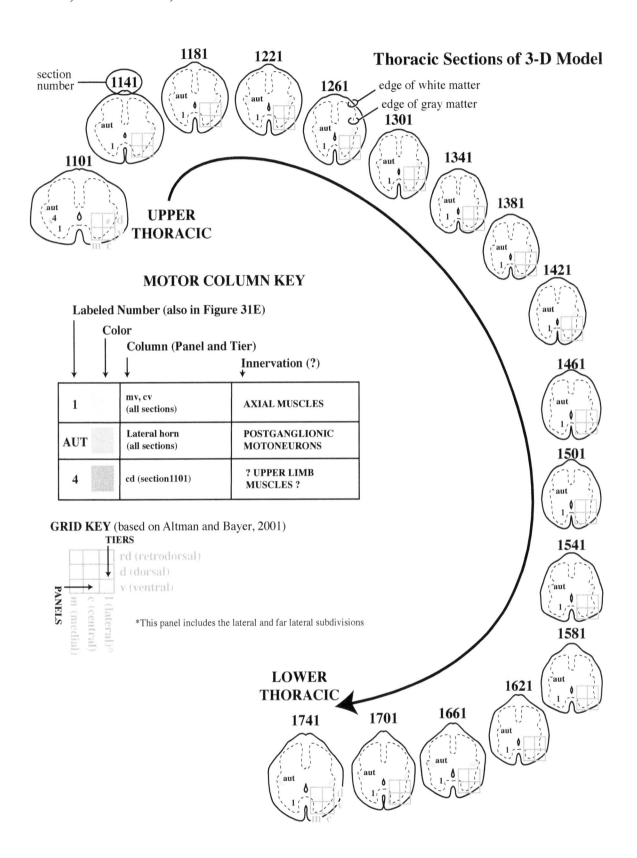

Thoracic Sections of 3-D Model

section number — 1141

edge of white matter
edge of gray matter

UPPER THORACIC

LOWER THORACIC

MOTOR COLUMN KEY

Labeled Number (also in Figure 31E)

Color

Column (Panel and Tier)

Innervation (?)

1		mv, cv (all sections)	AXIAL MUSCLES
AUT		Lateral horn (all sections)	POSTGANGLIONIC MOTONEURONS
4		cd (section1101)	? UPPER LIMB MUSCLES ?

GRID KEY (based on Altman and Bayer, 2001)

TIERS
rd (retrodorsal)
d (dorsal)
v (ventral)

PANELS
m (medial)
c (central)
l (lateral)*

*This panel includes the lateral and far lateral subdivisions

Figure 26. A continuation of the previous figure showing outlines of the sections used in the thoracic region of the spinal cord. Three of these sections are illustrated in Volume 15. **Section 1141** is in **Plate 32**; **section 1581** is in **Plate 33**; **section 1741** is in **Plate 34**. After fixation, the total length of the thoracic part of the model is 22.4 mm. The thoracic model is shown by itself in **Figure 31**.

126

FIGURE 27

GW 14, CR 108 mm, Y68-65

Lumbar and Sacral Sections of 3-D Model

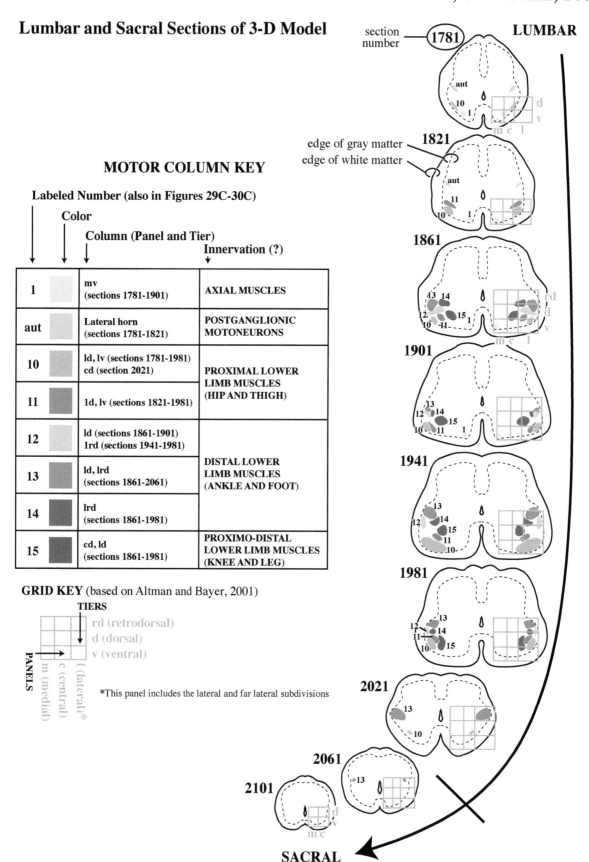

MOTOR COLUMN KEY

Labeled Number (also in Figures 29C-30C)

Color

Column (Panel and Tier)

Innervation (?)

Number	Color	Column (Panel and Tier)	Innervation (?)
1		mv (sections 1781-1901)	AXIAL MUSCLES
aut		Lateral horn (sections 1781-1821)	POSTGANGLIONIC MOTONEURONS
10		ld, lv (sections 1781-1981) cd (section 2021)	PROXIMAL LOWER LIMB MUSCLES (HIP AND THIGH)
11		ld, lv (sections 1821-1981)	
12		ld (sections 1861-1901) lrd (sections 1941-1981)	DISTAL LOWER LIMB MUSCLES (ANKLE AND FOOT)
13		ld, lrd (sections 1861-2061)	
14		lrd (sections 1861-1981)	
15		cd, ld (sections 1861-1981)	PROXIMO-DISTAL LOWER LIMB MUSCLES (KNEE AND LEG)

GRID KEY (based on Altman and Bayer, 2001)

TIERS
rd (retrodorsal)
d (dorsal)
v (ventral)

PANELS
m (medial)
c (central)
l (lateral)*

*This panel includes the lateral and far lateral subdivisions

Figure 27. A continuation of the previous figure showing outlines of the sections used in the lumbar (1781-1981) and sacral (2021-2101) regions of the spinal cord. **Section 1821** is in **Plate 35**; **section 1901** is in **Plate 36**, section **2061** is in **Plate 37**. After fixation, the total length of the lumbosacral part of the model is 11.2 mm. The lumbosacral part of the model is in **Figures 32-33**.

FIGURE 28

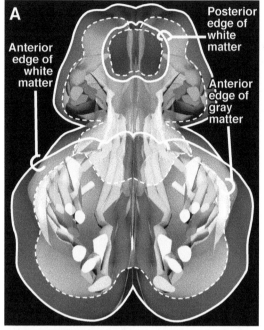

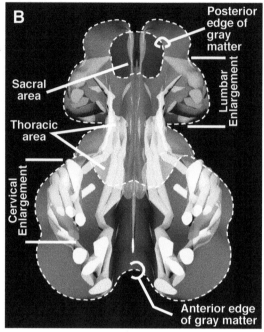

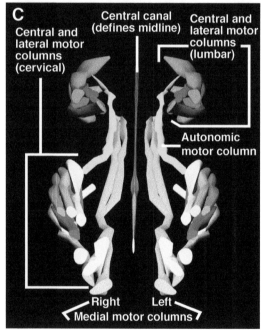

Figure 28

The entire spinal cord in specimen Y68-65 (GW14, CR 108 mm) as viewed from the front and top (**A, B, C,** this page) or the upper-right side (**D, E, F,** facing page). The front edge of the reconstruction is section 701, the back edge is section 2101. The total length after fixation is 49 mm. In this and the following reconstructions, the spinal cord has been enlarged 3 times more crosswise than lengthwise. Otherwise, the model is too long to see the various motoneuron columns clearly.

Solid outlines define the outer edges of the white matter (transparent envelope, **A** and **D**); **dashed outlines** define the outer edges of the gray matter (less transparent envelope, **A, B** and **D, E**). The **gray-white solid** in panels **A-E** is the central canal. The **colored solids** in panels **A-E** are ventral horn motor columns.

There are **15 pairs of motor columns** on each side of the ventral horn in this specimen. The medial motor columns (**cyan**) are split in the thoracic region and extend into the lumbar enlargement down to section 1901 and then disappear. The remaining motor columns are in the ventral, central, and lateral sectors. There are 8 central and lateral motor columns in the cervical region (**light brown, orange, yellow shades**). The lumbar enlargement is large in this specimen and contains 6 central and lateral motor columns (**purple, violet, magenta shades**). The top and side views in panels **C** and **F** show the intertwined arrangement of the motor columns in the cervical and lumbar enlargements. The autonomic motor columns (**green**) in the lateral horn of the thoracic and upper lumbar regions are also reconstructed in this specimen.

GW 14, CR 108 mm, Y68-65

FIGURE 28

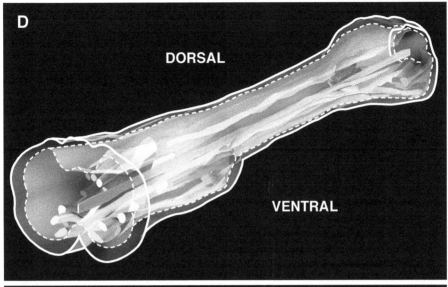

D

DORSAL

VENTRAL

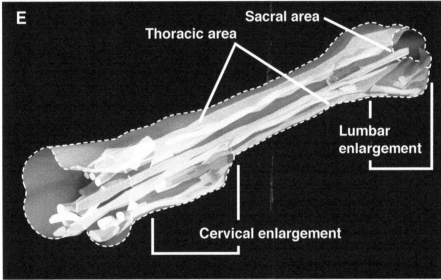

E

Thoracic area

Sacral area

Lumbar enlargement

Cervical enlargement

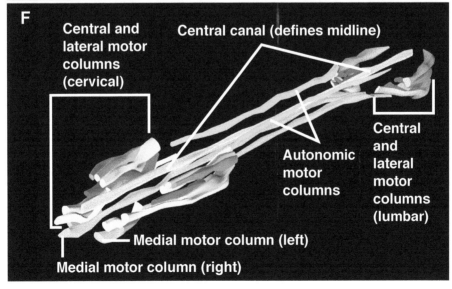

F

Central and lateral motor columns (cervical)

Central canal (defines midline)

Central and lateral motor columns (lumbar)

Autonomic motor columns

Medial motor column (left)

Medial motor column (right)

FIGURE 29

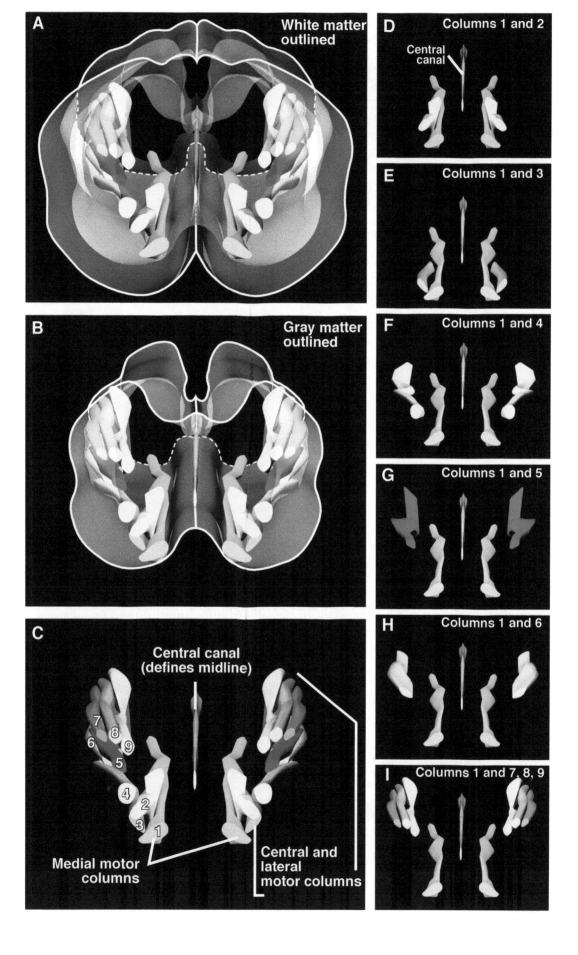

A White matter outlined

B Gray matter outlined

C Central canal (defines midline)

7
8
6 9
5
4
2
3 1

Medial motor columns

Central and lateral motor columns

D Columns 1 and 2

Central canal

E Columns 1 and 3

F Columns 1 and 4

G Columns 1 and 5

H Columns 1 and 6

I Columns 1 and 7, 8, 9

FIGURE 29

The top front view of the cervical region in specimen Y68-65 (GW 14, CR 108 mm). The individual sections (701-1061) in this model are diagrammed in **Figure 23**. The length of the cervical region is 12.6 mm after fixation.

All panels show the entire model or parts of it from the top front.

Panel **A** shows both the white matter (**outlined** transparent envelope) and gray matter (inner transparent envelope) around the central canal (**gray-white solid**) and motor columns (**colored solids**).

Panel **B** shows the only gray matter (**outlined** transparent envelope) around the central canal and motor columns.

Panel **C** shows the central canal and all motor columns alone. Subsets of motor columns are shown in the small side panels **D-I**.

FIGURE 30

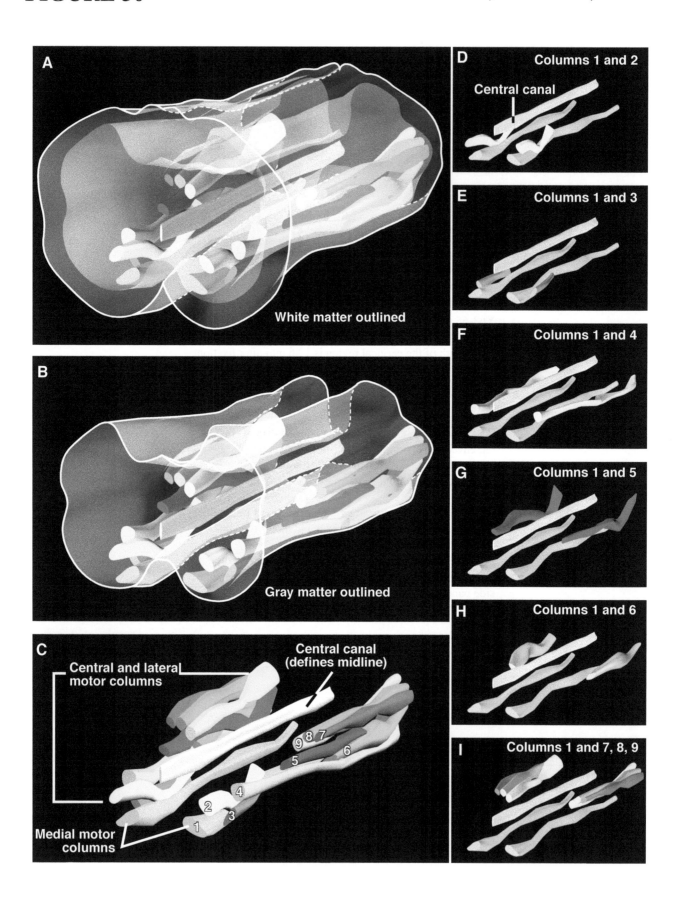

A

White matter outlined

B

Gray matter outlined

C

Central and lateral motor columns

Central canal (defines midline)

9 8 7
6
5
4
2
3
1

Medial motor columns

D Columns 1 and 2

Central canal

E Columns 1 and 3

F Columns 1 and 4

G Columns 1 and 5

H Columns 1 and 6

I Columns 1 and 7, 8, 9

FIGURE 30

An upper-right view of the cervical spinal cord in Y68-65.

There are **9 Motor Columns**:

1 (**cyan**, axial muscle motoneurons?) is located in the medial panel, ventral tier.

2-3 (**yellow, orange**) at upper cervical levels (sections 701-821) may innervate proximal upper limb muscles, and are located in the central and lateral panels, ventral and dorsal tiers.

4–6 (**yellow brown**, **red**, **orange**) are located in the cervical enlargement in the central and lateral panels, ventral, dorsal, and retrodorsal tiers and may innervate arm and forearm muscles.

7–9 (**dark red orange**, **light red orange**, **yellow orange**) are located in the cervical enlargement in the lateral panel, retrodorsal tier and may innervate wrist, hand, and digit muscles.

FIGURE 31

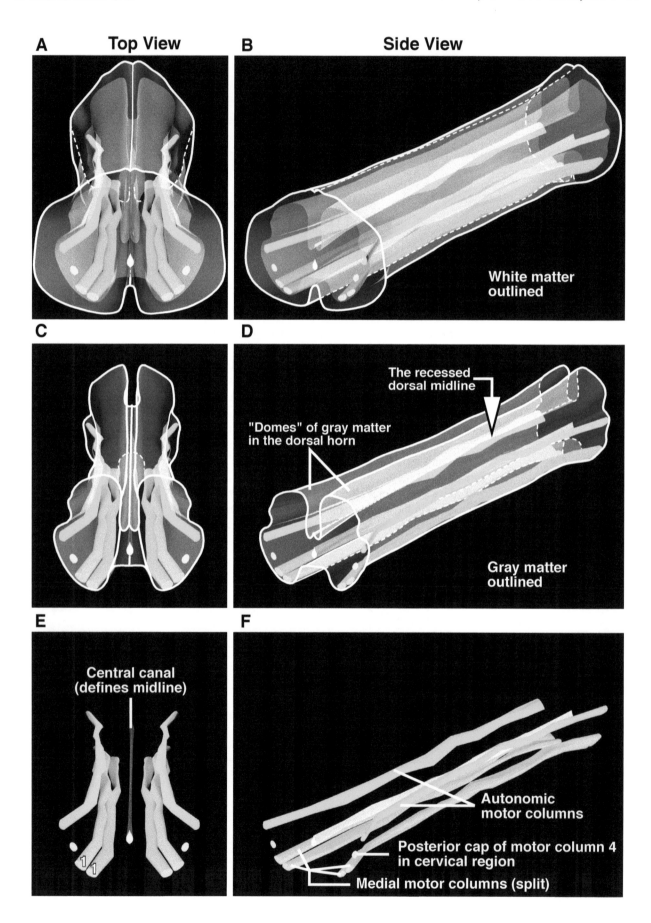

A   Top View

B   Side View

White matter outlined

C

D

The recessed dorsal midline

"Domes" of gray matter in the dorsal horn

Gray matter outlined

E

Central canal (defines midline)

F

Autonomic motor columns

Posterior cap of motor column 4 in cervical region

Medial motor columns (split)

FIGURE 31

The thoracic region in specimen Y68-65 (GW 14, CR 108 mm). The individual sections (1101-1741) in this model are diagrammed in **Figure 24**. The reconstructed length is 22.4 mm after fixation. Panels **A**, **C**, and **E** show the model from the top front, panels **B**, **D**, and **F** from the upper-right side. Panels **A** and **B** show both the white matter (**outlined** transparent envelope) and gray matter (inner transparent envelope) around the central canal (**gray-white solid**) and the medial motor columns (**1**, **cyan**), autonomic motor columns (**green**), and the posterior cap of motor column **4** (**yellow-brown**) from the cervical region that is present only in the first section (1101).

Panels **C** and **D** show the only gray matter (**outlined** transparent envelope) around the central canal and the motor columns.

Panels **E** and **F** show the central canal and the motor columns alone. This region of the spinal cord is distinguished from other regions by having a bifurcated motor column **1** that probably innervates axial muscles associated with the thoracic vertebrae, rib cage and dorsal abdominal wall. The autonomic motor columns in the lateral horn are prominent and innervate neurons in the sympathetic ganglia.

FIGURE 32

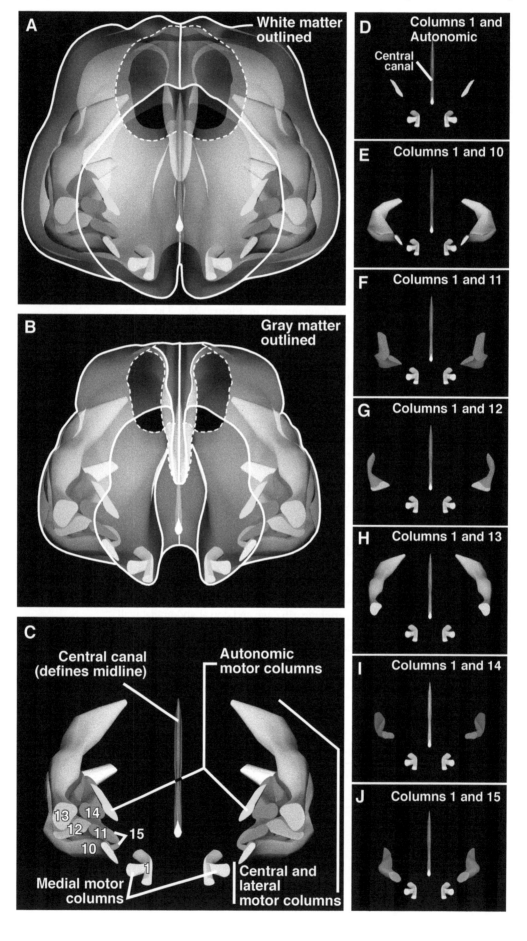

A — White matter outlined

B — Gray matter outlined

C — Central canal (defines midline) — Autonomic motor columns

13 14
12 11 →15
10
1

Medial motor columns — Central and lateral motor columns

D — Columns 1 and Autonomic — Central canal

E — Columns 1 and 10

F — Columns 1 and 11

G — Columns 1 and 12

H — Columns 1 and 13

I — Columns 1 and 14

J — Columns 1 and 15

FIGURE 32

The lumbar and sacral regions in specimen Y68-65 (GW 14, CR 108 mm). The individual sections (1781-2101) in this model are diagrammed in **Figure 25**. The reconstructed length is 11.2 mm after fixation. All panels show the model from the top front.

Panel **A** shows both the white matter (**outlined** transparent envelope) and gray matter (inner transparent envelope) around the central canal (**gray-white solid**) and motor columns (**colored solids**).

Panel **B** shows the gray matter (**outlined** transparent envelope) around the central canal and motor columns. Panel **C** shows the central canal and all the motor columns.

The small side panels (**D-J**) show various subsets of motor columns with the central canal.

FIGURE 33

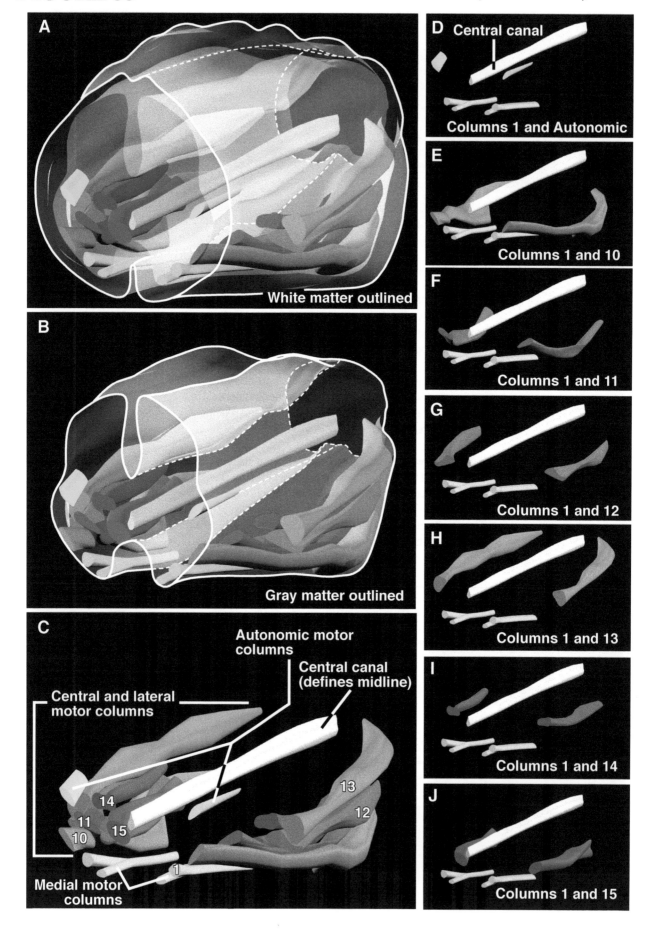

A White matter outlined

B Gray matter outlined

C Autonomic motor columns
Central and lateral motor columns
Central canal (defines midline)
14
13
11
10
15
12
1
Medial motor columns

D Central canal
Columns 1 and Autonomic

E Columns 1 and 10

F Columns 1 and 11

G Columns 1 and 12

H Columns 1 and 13

I Columns 1 and 14

J Columns 1 and 15

FIGURE 33

A continuation of the previous figure in the same specimen except that all panels show either the entire model or parts of it in a side view.

Motor column **1** (**cyan**, medial panel, ventral tier) probably innervates axial muscles. It is bifurcated in the first section (1781) and single in sections 1821 to 1861.

The autonomic motor column (**green**) is in the first two sections of the model (1781-1821) and contains the most posterior neurons that innervate the sympathetic ganglia.

The remaining motor columns (**10-15**) are present only at lumbar levels or upper sacral levels (sections 1781-2061).

Columns **10** (**pale violet**) and **11** (**dark violet**) may innervate proximal lower limb muscles controlling the hip joint and are located in the lateral panel, ventral and dorsal tiers.

Column **12**, **13**, and **14** (**pale violet**, **pink**, **purple**) are located in the lateral panel, dorsal and retrodorsal tiers. They may innervate lower limb muscles controlling the ankle and foot.

Column **15** (**reddish-magenta**) is located in the central and lateral panels, dorsal tier and may innervate leg muscles.